A Scotsman Returns

A Scotsman Returns

Travels with Thomas Telford in the Highlands and Islands

Paul A. Lynn

Whittles Publishing

Published by
Whittles Publishing Ltd.,
Dunbeath,
Caithness, KW6 6EG,
Scotland, UK

www.whittlespublishing.com

© 2021 Paul A. Lynn

ISBN 978-184995-486-0

By the same author

The Lighthouse on Skerryvore 978-184995-140-1

Scottish Lighthouse Pioneers:
Travels with the Stevensons in Orkney and Shetland 978-184995-265-1

World Heritage Canal:
Thomas Telford and the Pontcysyllte Aqueduct 978-184995-398-6

see https://www.whittlespublishing.com/Paul_A__Lynn

Printed in the UK by Cambrian Printers Ltd.

Contents

Preface

Two hundred years ago Thomas Telford (1757–1834), Scotland's greatest civil engineer, was busy completing a remarkable 20-year programme of work in the Highlands and Islands. This book follows in his footsteps – a modern travelogue and commentary on the infrastructure he designed and built, the places he visited, and the people he met.

Born in poverty and educated at a parish school, Telford developed a lifelong passion for self-education that led him from apprentice stonemason via budding architect to famous engineer. His work was very much in tune with the Scottish Enlightenment and had far-reaching social consequences. To put all this in context my book begins with an account of his early years in the Lowlands and Edinburgh, his middle years in England and Wales, and the state of Scotland as he found it on his return.

A few years ago I purchased a copy of *Journal of a Tour in Scotland in 1819* by Robert Southey, and it changed my perceptions of Thomas Telford. Southey, at that time the poet laureate, wrote a delightful account of an adventurous six weeks spent with Telford in the Highlands. The poet's interest in engineering and the engineer's love of poetry had brought them together, and they got on famously. Southey paints a picture of Telford very different from the rather dry reputation he has gathered over the years as a man totally devoted to his work. Instead he is revealed as a fully rounded character with a great sense of humour, a love of anecdote, and a talent for mixing with Highlanders both haughty and humble. A substantial part of my book is devoted to Southey's journal.

Later chapters cover the infrastructure built by Telford but not visited by Southey, and include some of my own reminiscences of memorable travels in Scotland over many years. I end with a brief assessment of Telford's Scottish legacy – a gift to his native land that must surely touch the hearts of all who love the Highlands and Islands.

Paul A. Lynn

A note on measurement units: the dimensions of the structures built by Telford and his contemporaries are given in feet, whereas the dimensions of more recent structures and those of natural features such as mountains are given in both feet and metres.

Acknowledgements and references

I am grateful to the following organisations and individuals for images which do so much to bring this story to life. All are acknowledged where they appear in the text.

Institution of Civil Engineers (ICE), London
Geograph Britain and Ireland
Wikipedia

The Geograph project, a charity set up in 2005, aims to collect geographically representative photographs and information of every square kilometre of Great Britain and Ireland. By early 2020 over 13,000 photographers had submitted about 6 million images covering 280,000 grid squares. The images are free to use providing due acknowledgement is given to Geograph and individual copyright holders. I have found this an invaluable resource, and wish to thank all the photographers whose images appear in the text. I would also like to thank Anne Burgess for the use of her article on Telford's Highland churches.

I have also found Wikipedia very useful for the information it contains and the leads it provides, and have drawn on these from time to time.

Among many books I have consulted, Anthony Burton's *Thomas Telford: Master Builder of Roads and Canals* has been particularly helpful. I should also mention L.T.C. (Tom) Rolt's classic books *Thomas Telford* (1958) and *Landscape with Canals* (1977), and I wish to thank Tim Rolt for permission to include a number of quotations from his father's inspirational writing.

Paul A. Lynn

Part 1

Precedents

Thomas Telford

A Scottish start

Among the remarkable engineers who made Britain's industrial revolution possible, Thomas Telford (1757–1834) is an outstanding example. Coming from the humblest of backgrounds, he taught himself the science and practice of civil engineering and went on to build marvellous canals, roads, bridges, and harbours. He never married, and devoted his time almost totally to his work. As a result he has sometimes been painted as a rather one-dimensional character who avoided the emotional peaks and troughs that accompany most lives. But to delve deeper is, as so often, to reveal a more complex tapestry.

This book takes the form of a modern travelogue, following in Telford's footsteps among the Highlands and Islands as he planned and executed one of the most remarkable programmes of civil infrastructure ever attempted in Britain. I hope to show him as a man of many parts – a lover of poetry and the natural world, a good companion, a compassionate man with an infectious sense of humour, an exemplar of the Scottish Enlightenment whose work had profound social consequences.

I suggest we start by visiting his Lowland birthplace, which reveals the enormous effect his roots had on his subsequent career and the pride he took in his native land right up to the day of his death in London.

Thomas Telford entered the world in a shepherd's cottage in Glendinning, a stone's throw from the hamlet of Jamestown, about 20 miles north of Gretna Green and 8 miles from Langholm, the local town. Apart from recent afforestation, the countryside of his childhood has changed little, and remains enticingly far from the madding crowd. Thankfully, however, it is now served by good roads and we may get a good feel for it by travelling from Langholm along the B709 towards Eskdalemuir. This is gentle country by Scottish standards, more rural idyll than Highland drama, with rich green hills sweeping down to valley floors and sparkling streams feeding the River Esk.

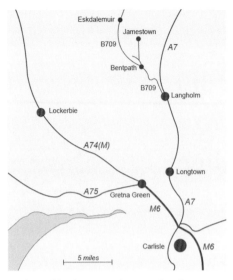

The Telford cottage lay in Glendinning, a stone's throw from the hamlet of Jamestown (top centre).

After 5 miles we come to Bentpath, larger than a hamlet, smaller than a village, in the expansive rural parish of Westerkirk. But rather than turn right to Westerkirk Bridge and the church, we will follow the B road a little further to discover a surprising building for such a spot: neither chapel nor village school, but a *library*; and one that has been described as the best equipped, given its location and the population it serves, of any in the country. It is also the oldest library still in use in Scotland, and holds over 8,000 books dating from the 18th century onwards, with records that give a fascinating insight into the reading habits of a small rural community. There must be a good story here somewhere.

That story dates from 1791 when the Westerhall Mining Company started mining and smelting antimony at Glendinning's Louisa Mine, just a stone's throw from the cottage where Thomas Telford had been born 34 years previously. The company built an access road and bridges, and a row of miners' cottages beside Meggat Water; and much more surprisingly the library, which completed the settlement of Jamestown. But the mine had a short life, closing in 1799 after yielding 100 tons of semi-pure antimony with a value of £8,400. The library was relocated in the Bentpath village school that Thomas had attended, and again in 1840 to a brand new building – the one we see today.

As we shall discover, Thomas became fascinated by books in his teenage years and carried a love of poetry through to the end of his life. It is doubtful whether he ever saw miners at work in the short-lived Louisa Mine because by the 1790s he was engaged far away as a professional engineer, but he must have heard about the library and its subsequent removal to cramped conditions in the village school. When he died in 1834 he bequeathed £1,000 'to the minister of Westerkirk in trust for the parish library', a princely sum which had grown to £2,700 by the time his estate was settled.

Today we see the fine Victorian building funded by Thomas Telford, and a more recent stone memorial at the roadside. One of its plaques announces:

> This seat was erected in 1928 to perpetuate the memory of Thomas Telford, son of the unblameable shepherd, and to record his fame as an engineer and his untiring benevolence. Apprenticed to a stonemason in Langholm, his creative genius gave to the nation many works of inestimable benefit. He was the first President of the Institution of Civil Engineers.

Another plaque carries lines by the man himself:

Westerkirk Library and the Telford memorial seat near Bentpath (Geograph/Walter Baxter).

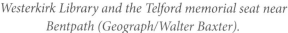

> *There 'mongst those rocks I'll form a rural seat*
> *and plant some ivy with its moss compleat*
> *I'll benches form of fragments from the stone*
> *which nicely pois'd was by our hands oerthrown.*

By the early 1990s the library was in urgent need of some TLC. An ambitious five-year project restored the building and its stock of books to their former glory, giving visitors a glimpse of records going back over 200 years. At the time of writing admission is free but opening times are limited. The borrowing of books is restricted to residents of Westerkirk and the adjoining parishes.

We now retrace our steps a few hundred yards back along the B709 and turn into the lane that leads over Westerkirk Bridge to Bentpath. The bridge, built some 20 years before Thomas's birth, overlooks a bustling river. He must have wandered over it hundreds of times as a child and teenager, little dreaming that he would become one of the world's great bridge builders. Nearby is Westerkirk parish church (not the building he knew) and the churchyard with a simple headstone that marks the grave of a poor shepherd, embellished with the earliest known example of letter carving by a young man who would soon become a highly skilled stonemason:

> IN MEMORY OF
> JOHN TELFORD,
> WHO AFTER LIVING 33 YEARS
> AN UNBLAMEABLE SHEPHERD,
> DIED AT GLENDINNING,
> NOVEMBER, 1757.

The shepherd was, of course, his father.

Westerkirk Bridge over the River Esk, completed in 1736 (Geograph/Walter Baxter).

Almost subconsciously I have started to call the baby and the young lad Tom. Previously, like most of his biographers and commentators, I preferred Thomas Telford, or just plain Telford. Yet we are about to discover a birthplace so humble, and a family situation so fraught, that a description of his childhood seems best uncluttered by adult formality. Time enough later for Telford, giant of the civil engineering profession.

The lane bears left at the far end of Westerkirk Bridge to follow the north bank of the Esk. A mile or so later a turning on the right leads up alongside Meggat Water to Jamestown; and finally we reach Glendinning and a simple stone cairn, erected in 2007 on a knoll where a mud-walled cottage 'little better than a shieling' once stood – the spot where Tom was born among the hills.

This is a good moment to introduce Samuel Smiles (1812–1904), a famous Victorian author who had a clear view of Tom's humble beginnings in Eskdale:

> One would scarcely have expected to find the birthplace of the builder of the Menai Bridge and other great national works in so obscure a corner of the kingdom. Possibly it may already have struck the reader with surprise, that not only were all the early engineers self-taught in their profession, but they were brought up mostly in remote country places, far from the active life of great towns and cities. But genius is of no locality, and springs alike from the farmhouse, the peasant's hut, or the herd's shieling. Strange, indeed, it is that the men who have built our bridges, docks, lighthouses, canals, and railways, should nearly all have been country-bred boys.

The Telford cairn at Glendinning (Geograph/Walter Baxter).

His description of Glendinning and the surrounding countryside shows that they have changed remarkably little over the years:

> From the knoll may be seen miles on miles of hills up and down the valley, winding in and out, sometimes branching off into smaller glens, each with its gurgling rivulet of peaty-brown water flowing down from the mosses above … at Glendinning you seem to have got almost to the world's end. There the road ceases, and above it stretch trackless moors, the solitude of which is broken only by the whimpling sound of the burns on their way to the valley below, the hum of bees gathering honey among the heather, the whirr of a blackcock on the wing, the plaintive cry of ewes at lambing-time, or the sharp bark of the shepherd's dog gathering the flock together for the fauld.

But in 1757 the rural idyll was accompanied by tragedy. Tom's mother had lost a previous baby Tom, and the happy arrival of a successor was overshadowed by dreadful news, further reported by Smiles:

> In this cottage on the knoll Thomas Telford was born on the 9th of August, 1757, and before the year was out he was already an orphan. The shepherd, his father, died in the month of November, and was buried in Westerkirk churchyard, leaving behind him his widow and her only child altogether unprovided for.

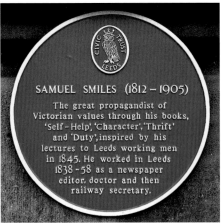

Samuel Smiles, a man with many interests (Wikipedia).

We will do well to follow Samuel Smiles from now on, because he wrote admirably about Thomas Telford in his famous five-volume series 'Lives of the Engineers' published in London in 1862. Born in East Lothian, Scotland, one of 11 children, Smiles studied medicine at Edinburgh University. He went on to develop an unlikely passion for political journalism, especially as it related to social and economic conditions in Victorian Britain. By the age of 30 he was editor of the *Leeds Times* and secretary to the Leeds Parliamentary Reform Association, a Chartist organisation with an agenda that for its time was little short of revolutionary. Three years later he branched out again to become secretary of the newly formed Leeds & Thirsk Railway, and subsequently of the South Eastern Railway. By this time his passion for parliamentary reform was displaced by a growing conviction that individual self-improvement was a surer remedy for society's many ills. And when in 1859, at the age of 47, his book *Self-Help* was published, he became a national celebrity.

The origins of *Self-Help* lay in a speech he had given to a mutual improvement society, claiming that poverty was largely caused by irresponsible personal behaviour. Social progress would come from new attitudes rather than new laws. However, lest this be thought unduly harsh and against his Chartist sympathies, he also attacked rampant materialism and the injustices of laissez-faire capitalism. The book has been called 'the bible of mid-Victorian liberalism'.

Lives of the Engineers appeared two years later. It seems that Smiles' experience as secretary of two new railway companies had stimulated an interest in the engines of the industrial revolution and, more importantly from our point of view, in the remarkable characters who made it possible. An author veering towards the primacy of self-improvement over government action for the progress of society would have found plenty to applaud in the great engineers who in most cases reached the top of their profession without any privileges of wealth or family background.

Back in Eskdale, Tom's mother, from now on to be known by her maiden name, Janet Jackson, in line with local custom, was in deep trouble: almost penniless, living in a tied cottage that must be surrendered now that her 'unblameable' husband had passed away, alone with an infant of four months to be nursed, fed, and clothed. She would face a long hard struggle with enormous fortitude, overcoming problems that would have destroyed a lesser spirit. It seems that baby Tom, destined to become an engineer of legendary ambition and determination, inherited a generous assortment of genes from his mother.

Fortunately for Janet and her son, the lawlessness that had existed for centuries in the border country between Scotland and England, where 'reivers' rustled livestock, stole household goods, took prisoners for ransom, and even murdered, had subsided since the Act of Union in 1707. Extended families, referred to as clans, who had acted as miniature armies and devoted their energies to plunder, were now in more cooperative mood. The Esk valley was among the beneficiaries, and although poverty was still rife, there was less fear and misery. As Smiles noted:

> The farmers of the dale were very primitive in their manners and habits, and being a warm-hearted, though by no means a demonstrative race, they were kind to the widow and her fatherless boy. They took him by turns to live with them at their houses, and gave his mother occasional employment. In summer she milked the ewes and made hay, and in harvest she went a-shearing; contriving not only to live, but to be cheerful.

A raid by reivers on Gilnockie Tower, Dumfries and Galloway (Wikipedia/Cattermole).

So in spite of a total absence of state-funded social security and medical care, Janet survived. In such a small community little or no stigma attached to accepting a neighbour's help, considered a friendly act that elevated the donor as well as the recipient. She was soon moved about a mile down the valley to a place called The Crooks, and found herself rehoused in another primitive mud-walled thatched cottage, this time with a central doorway and two rooms, one for her and Tom, the other for neighbours called Elliot. From there she brought up an only child who, as an old man greatly honoured in his own lifetime, would declare 'I still recollect with pride and pleasure my native parish of Westerkirk, on the banks of the Esk, where I was born.'

The child grew into a healthy boy, so full of fun that he became known in the valley as Laughing Tam. When he was old enough to herd sheep he stayed with a relative, a shepherd like his father, and spent much of his time on the hills. In winter he often lodged with neighbouring farmers, herding cows or running errands and receiving in return food, a pair of stockings, and wooden clogs.

Time spent alone among the sweeping hills of Eskdale had a profound effect on young Tom, encouraging a deep love of the natural world and his native place. It is surely no coincidence that as an eminent engineer he would execute many of his greatest works in wild areas of the Scottish Highlands and North Wales – works that show a great sensitivity to landscape and the grandeur of their setting.

Education was also on the agenda. Westerkirk was fortunate to have a parish school, an admirable institution that developed at an early date in Scotland. The modest fees were beyond Janet's means so she reluctantly accepted help from her brother – help that proved of enormous benefit:

> By imparting the rudiments of knowledge to all, the parish schools of the country placed the children of the peasantry on a more equal footing with the children of the rich; and to that extent addressed the inequalities of fortune. To start a poor boy on the road of life without instruction, is like starting one on a race with his eyes bandaged or his leg tied up … It was not much that he learnt … but there was another manifest advantage to the poor boy in mixing freely at the parish school with the sons of the neighbouring farmers and proprietors. Such intercourse has an influence on a youth's temper, manners, and tastes, which is quite as important in the education of character as the lessons of the master himself; and Telford often, in after life, referred with pleasure to the benefits which he had derived from his early school friendships.

Prominent among Thomas's friends was Andrew Little, who went on to study medicine in Edinburgh and later enrolled as ship's doctor on an African slaving vessel. The voyage was a disaster, and not only for the slaves, because lightning savaged the ship during a storm and left the young doctor permanently blinded. His career in medicine cruelly truncated, he returned to Langholm and became the village schoolmaster. When Thomas subsequently left Eskdale for London, he started a long correspondence with his blind friend and was always delighted to reminisce about their schooldays, joking about 'the glutton boy' John Elliot, and enquiring about Jennie Smith who it seems had taken his fancy – 'tell her she is a canterrin sort of lassie'.

But what would Tom do on leaving school? The 15-year-old son of a single mother living in a thatched mud cottage in a remote valley had few strings to pull, and the prospects for gainful employment must have looked bleak. Was he to become a shepherd like his father, or a farm labourer, or perhaps a trade apprentice? Local trade was largely confined to building stone walls, using skills that 'an ordinarily neat-handed labourer could manage' – not an inspiring prospect – and jobs in the rest of Eskdale were few and far between.

When Samuel Smiles visited the area many years later he asked a local what would happen to the many children he saw in local villages, and received a blunt answer:

> They swarm off! If they remained at home, we should all be sunk in poverty, scrambling with each other amongst these hills for a bare living. But our peasant-ry have a spirit above that: they will not be content to sink; they look up; and our parish schools give them a power of making their way in the world, each man for himself. So they swarm off – some to America, some to Australia, some to India, and some, like Telford, work their way across the border and up to London.

The quotation confirms that this member of the local 'peasantry' had sufficient spirit to 'swarm off'. But it jumps the gun on our story because another nine years would pass before Thomas Telford took the road to London. His flight from Eskdale was to take place in several distinct hops.

The first was to the village of Lochmaben, about 20 miles from home, where he started an apprenticeship with a local stonemason in 1772. But his new master ill-treated him – we have no details – and a few months later Laughing Tam was back with his mother at The Crooks. She must have been worried and disappointed, but luckily her nephew Thomas Jackson, who had recently been appointed land steward to the Johnstones, a wealthy local family, stepped in to help. As a man with some influence Jackson was able to persuade Andrew Thomson, a well-respected mason in Langholm, to take Thomas for the remainder of his apprenticeship.

So Thomas's second hop was to Langholm, and it benefited the future civil engineer enormously. Today's Langholm is a small neat town of about 2,500 people, well known for its woollen textile industry and as the historic centre of the Armstrong clan that made its own notable contribution to cross-border raids before the Act of Union. But back in 1772 it was a village consisting chiefly of 'mud hovels' and 'no better than the district that surrounded it'. Fortunately it was about to undergo major improvements, thanks to the young Duke of Buccleuch who had succeeded to his family's local estates five years previously. The Buccleuchs, like the Armstrongs, were a prime example of a family that had thrown its energy into border raids for centuries, but were now redirecting it, somewhat tardily, to less destructive activities:

> The energy which the old borderers threw into their feuds has not become extinct, but survives under more benignant aspects, exhibiting itself in efforts to

enlighten, fertilize, and enrich the country which their wasteful ardour before did so much to disturb and impoverish. The heads of the Buccleugh [*sic*] and Elliot family now sit in the British House of Lords … The border men, who used to make such furious raids and forays, have now come to regard each other, across the imaginary line which divides them, as friends and neighbours; and they meet as competitors for victory only at agricultural meetings, where they strive to win prizes for the biggest turnips or the most effective reaping machines.

It is clear that Thomas, who had entered the world with almost no prospect of advancement, now found himself in an environment that was changing rapidly for the better. In this he was lucky: had he been born 50 years earlier, the seeds of his ambition might well have fallen on stony ground; 50 years later, on fertile soil already claimed by other talented individuals. As it was, the Duke of Buccleuch's programme of works in Langholm – paving rough tracks, bridging fords, building stone houses in place of mud and thatch – produced a demand for masons' work at just the right time for Thomas, and introduced him to a wide variety of skills under his new master, Andrew Thomson.

Langholm had grown up on the east bank of the River Esk, but by the time Thomas had completed his apprenticeship it was extending westwards. He helped build houses in its New Town, some with ornamental doorways that demonstrated his increasing confidence as a craftsman in stone and the pride he took in his work. But above all it was a new bridge thrown over the Esk by a master mason, Robin Hotson, in cooperation with Andrew Thomson that introduced Thomas to the massive structures that would play such a big part in his professional life. Many of its stones were hewn by his hand and carried his personal mason's mark – a vertical line with an arrowhead at the top and a lozenge at the bottom, separated by a cross. When the Esk runs low in summer some of the marks may still be seen below the western arch.

Soon after the bridge was completed in 1778 an unusually high flood surged down the Esk valley. The river was 'roaring red frae bank to brae' and onlookers feared that the structure would be swept away. Hotson was away from home at the time and his wife Tibby, knowing he was under contract to maintain the bridge for seven years, panicked. She ran sobbing from one person to another, calling out 'Where's Tammy Telfer, where's Tammy? … they say it's shakin, it'll be doon.' In spite of Tammy's reassurances and evident amusement, she insisted the bridge was shifting and ran up to set her back against the parapet, to stiffen it. The bridge stands firm to this day.

Thomas worked in Langholm and Eskdale for eight years, latterly as a journeyman mason in his own right. He built farmhouses on the Duke of Buccleuch's estate and the Manse at Westerkirk. As work began to slacken off in Langholm he undertook small jobs including the hewing of gravestones and ornamental doorways. Throughout this period he kept in touch with his mother and often visited her at The Crooks on Saturday evenings, accompanied her to the parish church on Sundays, and displayed a compassion that would endure long after he became famous. But he had now learned all that his native valley could

Langholm Bridge over the River Esk, worked on by Thomas Telford. Old Langholm is on the right, New Town on the left (Geograph/John Chroston).

teach him in the art of masonry. In 1780 he decided to leave Eskdale and seek work 50 miles away in Edinburgh.

Thomas's years in Eskdale had introduced him to arts other than stonemasonry, and none more significant for his personal development than the riches of English literature. This had come about in a surprising way. An elderly lady, Miss Pasley, who lived in one of the few smart houses in Langholm, became aware of his family situation:

> As the town was so small that everybody in it knew everybody else, the ruddy-cheeked, laughing mason's apprentice soon became generally known to all the townspeople, and amongst others to Miss Pasley. When she heard that he was the poor orphan boy from up the valley, the son of the hard-working widow woman, Janet Jackson, her heart warmed to the mason's apprentice, and she sent for him to her house. That was a proud day for Tom; and when he called upon her, he was not more pleased with Miss Pasley's kindness than delighted at the sight of her little library of books, which contained more volumes than he had ever seen before.

He had already borrowed, and exhausted, the few books owned by his young friends, so he was delighted when Miss Pasley offered him volumes from her own collection. From now on he was almost never without a book, to be read during snatched intervals in his working day among his beloved hills, or on winter evenings by the uncertain light of a candle or cottage fire.

On one occasion Miss Pasley lent him Milton's *Paradise Lost*. Its 10,000 lines of verse covering the greatest of all themes, the struggle between good and evil, intoxicated him: 'I read and read, and glowred; then read and read again.' In a young man of 20 who was spending most of his time fashioning stone in the company of people with little formal education, this sensitivity to the rhythmic power of John Milton's epic poem seems remarkable. It probably stirred feelings experienced among the wilds of nature, but never so far expressed.

Thomas went on to read contemporary poets including Robert Southey (1774–1843), who became poet laureate and, as we shall see, accompanied him on a tour of the Scottish Highlands many years later. He also began to write verse of his own – usually, as he admitted, not very well: 'It is, to me, something like what a fiddle is to others, I apply to it in order to relieve the mind after being much fatigued with close attention to business.' Nevertheless some of his efforts, including lines addressed to his hero Robert Burns, and verses in praise of the hill country where he was born, achieved publication. The most ambitious of his early efforts was 'Eskdale' which prompted Southey to declare that 'many poems which evinced less observation, less feeling, and were in all respects of less promise, have obtained university prizes'.

Discovery of poetry and literature awakened Thomas to a wider world of ideas and helped him feel at ease in educated company. He became such a competent writer that he was often asked by acquaintances to pen letters on their behalf (this quote and the following three from Smiles):

> One evening a Langholm man asked Tom to write a letter for him to his son in England; and when the young scribe read over what had been written to the old man's dictation, the latter, at the end of almost every sentence, exclaimed, "Capital!, capital!" and at the close he said, "Well! I declare, Tom! Werricht himsel' couldna ha' written a better!" – Wright being a well-known lawyer or "writer" in Langholm.

Thomas's 1780 hop to Edinburgh introduced him to a New Town vastly more impressive than Langholm's. Princes Street was 'rising as if by magic' and skilled masons were in great demand. He arrived at a time of transformation, not only in city layout and architecture, but in science, medicine, industry, agriculture, political economy and philosophy, the fruits of Scotland's gathering Age of Enlightenment. New ideas based on evidence and reasoned argument were challenging traditional authority, especially religious authority, in the optimistic belief that humans could improve society without divine help – a philosophical stance similar in our time to that of Humanism. Its intellectual climate probably suited Thomas Telford's personality down to the ground.

He remained in Edinburgh for about 18 months

> during which he had the advantage of taking part in first-rate work and maintaining himself comfortably, while he devoted much of his spare time to drawing, in its application to architecture. He took the opportunity of visiting

Feasts for Thomas Telford's eyes: above l–r, Edinburgh Castle (Geograph/Richard Croft) and Holyrood House (Geograph/John Lord); below l–r, Heriot's Hospital, Edinburgh (Wikipedia/Oliver-Bonjoch) and Melrose Abbey (Geograph/James Barton).

and carefully studying the fine specimens of ancient work at Holyrood House and Chapel, the Castle, Heriot's Hospital, and the numerous curious illustrations of middle age domestic architecture with which the Old Town abounds.

Thomas was clearly aiming high – indeed, so high that even a buoyant Edinburgh could not contain him:

Having acquired the rudiments of my profession, I considered that my native country afforded few opportunities of exercising it to any extent, and therefore judged it advisable (like many of my countrymen) to proceed southward, where industry might find more employment and be better remunerated.

After a final architectural visit to Melrose Abbey, a former Cistercian monastery in border country about 40 miles southeast of Edinburgh, he returned to Eskdale to say goodbye to his old contacts, including the neighbouring farmers, who had befriended him and his mother when struggling with poverty – his schoolfellows, many of whom were preparing to migrate, like himself, from their native valley – and the many friends and acquaintances he had made while working as a mason in Langholm. Everybody knew that Tom was going south, and all wished him God speed.

Yet it seems surprising that Edinburgh's burgeoning New Town failed to satisfy his ambition, at least for a few more years, because it was becoming a showcase for the neoclassical architecture in vogue in England. The city fathers, realising that the hopelessly overcrowded Old Town no longer suited its growing professional and merchant classes, sanctioned the building of the first phase of New Town in 1767, 13 years before Thomas's arrival. The building works were to last until 1820. It is hard to believe that the talented stonemason exhausted Edinburgh's possibilities so rapidly.

Actually we do not really know the details of Thomas's stay in Edinburgh. He himself was silent on the subject. Samuel Smiles offers no clues. Perhaps the 24-year-old encountered personal difficulties with bosses or colleagues – or had met, and fallen for, a young woman who jilted him. In any case his mind was made up, and late in 1781 he embarked on his next adventure. Like Dick Whittington, but without the cat, he set off for London.

To England and Wales

Our Victorian author, Samuel Smiles, again sets the scene:

> A common working man, whose sole property consisted in his mallet and chisels, his leathern apron and his industry, might not seem to amount to much in "the great world of London". But, as Telford afterwards used to say, very much depends on whether the man has got a head with brains in it of the right sort upon his shoulders. In London, the weak man is simply a unit added to the vast floating crowd, and may be driven hither and thither, if he do not sink altogether; while the strong man will strike out, keep his head above water, and make a course for himself, as Telford did.
>
> There is indeed a wonderful impartiality about London. When work of importance is required, nobody cares to ask where the man who can do it best comes from, or what he has been, but what he is, and what he can do. Nor did it ever stand in Telford's way that his father had been a poor shepherd in Eskdale, and that he himself had begun his London career by working for weekly wages with a mallet and chisel.

So how did Thomas Telford and his hand tools get to the capital? His early years among the hills of Eskdale had made him a great walker and he thought nothing of 20 miles a day – but it was 350 miles to London. He had learned 'rough riding' as a youth, but did not own a horse. Travel by stage coach was certainly improving in the late 1700s, but the journey from Edinburgh to London meant staying in coaching inns and was beyond the means of a journeyman mason. None of these fitted the bill.

By chance his cousin Thomas Jackson came to the rescue once more. Jackson's employer, Sir James Johnstone of Wester Hall, had promised a horse to a family member in London but was unable to find a suitable rider. Not only did Jackson secure the invitation, but he lent Thomas Telford his buckskin breeches to cushion him against the hardships of the road. So the young mason left for London well mounted, and delivered Sir James's horse without incident. Jackson often told the story of his cousin's first ride to London with considerable relish, taking care to wind up with 'but Tam forgot to send me back my breeks!'

Thomas's luck continued. Before leaving Langholm he had been asked by his elderly friend Miss Pasley to deliver a letter to her brother John, a wealthy London merchant. As a result the young stonemason obtained an introduction to Sir William Chambers, architect of one of London's most famous neoclassical buildings, Somerset House, then under construction. It was 'the finest architectural work in progress in the metropolis' and Thomas wanted to be part of it. From the start he was determined to demonstrate his skills in fine carving rather than the mere fashioning of solid blocks, and was soon being employed as a first-class mason. He landed on his feet and, to judge by letters sent to friends in Eskdale, was happy, cheerful, and keen to receive news of those he had left behind. The gentler side of his personality surfaced in a letter written after more than a year's absence, envying the planned visit to Eskdale by a young surgeon acquaintance – 'for the meeting of long absent friends is a pleasure to be equalled by few other enjoyments here below'.

He worked with many fellow masons during his time at Somerset House and his Laughing Tam personality stood him in good stead. Yet quiet observation convinced him that most of his companions lacked spirit and, above all, forethought:

> He found very clever workmen about him with no idea whatever beyond their week's wages. For these they would make every effort: they would work hard, exert themselves to keep their earnings up to the highest point, and very readily "strike" to secure an advance; but as for making a provision for the next week, or the next year, he thought them exceedingly thoughtless. On the Monday mornings they began "clean", and on Saturdays their week's earnings were spent.

Somerset House, London, soon after completion, showing its façade onto the Strand and its frontage onto the River Thames (Wikipedia).

Unsurprisingly, his own approach was entirely different:

> Telford, on the other hand, looked upon the week as only one of the storeys of a building; and upon the succession of weeks, running on through years, he thought that the complete life structure should be built up.

Yet again we get the impression of a young man who, despite the humblest of beginnings, harboured an extraordinary ambition; every work experience must be put to good use, every shilling earned was means to an end, not an end in itself.

The most talented of his fellow masons, a man called Hatton, was 'honesty and good nature itself' and the only one to become a close friend. He had been working on Somerset House for six years and was 'esteemed the finest workman in London, and consequently in England'. With a thorough understanding of architectural drawings, he worked equally well in stone and marble, cutting Corinthian capitals and other elaborate ornaments with confidence. But he, too, was content to work as a journeyman, earning just a few shillings a week above average wages.

Telford suggested to his friend that they set up in business together, noting that there was 'nothing done in stone or marble that we cannot do in the completest manner'. Robert Adam, one of the most famous neoclassical architects of the age and an enthusiast of the Grand Tour, learned of their plans and promised support. But it was not to be: their great difficulty was lack of business capital. Thomas eventually had to accept it as an insuperable hurdle and with great regret abandoned the scheme; but it was an early indication of a growing determination to become his own master.

Perhaps that is what led to Thomas's next major hop. In 1784, at the age of 27, he obtained a contract to build a fine residence for the commissioner of Portsmouth Naval Dockyard, designed by another well-known neoclassical architect, Samuel Wyatt. The contacts Thomas had made in London, and especially his work at Somerset House, were bearing fruit. The contract included several other buildings and a new chapel (now St Ann's Church), giving him the chance to roam among the dockyard's many facilities, including the massive Great Stone

Portsmouth Naval Dockyard: the Commissioner's House,
built by Telford; and the Great Stone Dock (Wikipedia).

Dock of 1689, which had recently been upgraded. It is also likely that he saw HMS *Victory*, either berthed in the dockyard or sailing close offshore in Spithead. Commissioned in 1778, she was destined to become Nelson's flagship at Trafalgar in 1805. Telford must have been aware of her growing reputation as a formidable first-rate ship of the line, and inspired by yet another example of Britain's growing industrial and naval pre-eminence.

Thomas Telford's buildings progressed so satisfactorily that the commissioner started to value his advice above that of the dockyard superintendent – 'a dangerous point, being difficult to keep their good graces as well as his. However I will contrive to manage it.' His daily schedule was tightly controlled: rising at 7 in the morning in winter, 5 in summer, he worked on business matters until breakfast at 9; spent the rest of the morning among his buildings, checking progress and giving orders; lunched at 2, followed by writing, drawing, and reading until 9.30; then supper and bed.

> This is my ordinary round, unless when I dine or spend an evening with a friend; but I do not make many friends, being very particular, nay, nice to a degree. My business requires a great deal of writing and drawing ... Then, as knowledge is my most ardent pursuit, a thousand things occur which call for investigation which would pass unnoticed by those who are content to trudge only in the beaten path. I am not contented unless I can give a reason for every particular method of practice which is pursued. Hence I am now very deep in chemistry ... I have borrowed a MS copy of Dr Black's lectures. I have bought his "Experiments on Magnesia and Quicklime", and also Fourcroy's Lectures, translated from the French by one Mr Elliot, of Edinburgh. And I am determined to study the subject with unwearied attention until I attain some accurate knowledge of chemistry, which is of no less use in the practice of the arts than it is in that of medicine.

Such words are crucial for our understanding of Thomas Telford, whose formal education had stopped when he left the parish school in Westerkirk. He had received no help from high school or university; now, at the age of 27, he was immersing himself in the mysteries of chemistry, a prelude to a lifelong course of self-education. The following year he wrote to his blind friend Andrew Little in Langholm that he was 'chemistry mad', and that the pocket book which he always carried was becoming crammed with 'facts relating to mechanics, hydrostatics, pneumatics, and all manner of stuff'.

But in case we should think him a narrow workaholic, he assures us that he is taking a great delight in Freemasonry, and is about to have a room decorated and furnished to his own design at the George Inn, where he has his hair powdered every day and puts on a clean shirt three times a week. Not that attention to personal appearance denotes conceit – he would 'rather have it said of him that he possessed one grain of good nature or good sense than shine the finest puppet in Christendom'. And no matter how much he was burdened with work, he continued to write letters home, especially to his mother in The Crooks, using block capitals so that she could decipher them by her cottage fireside. As Samuel Smiles remarks:

As a man's real disposition usually displays itself most strikingly in small matters – like light, which gleams the most brightly when seen through narrow chinks – it will probably be admitted that this trait, trifling though it may appear, was truly characteristic of the simple and affectionate nature of the hero of our story.

Thomas completed his buildings in Portsmouth Dockyard in late 1786, and his contract came to an end. There was not much else to detain him in Portsmouth, but where should he go next – back to London, where neoclassical architecture was in full flow; or perhaps Edinburgh, where his newly acquired knowledge and experience of prestigious buildings would surely open new doors? But as it happened his next opportunity was offered by another contact from the past: William Pulteney, who had moved away from Eskdale many years before and was by now a highly influential member of Parliament for Shrewsbury, county town of Shropshire.

Pulteney was the second son of Sir James Johnstone of Wester Hall in Eskdale, the man who had taken on Thomas's cousin as land steward and entrusted one of his horses to Thomas for his ride to London. Pulteney had studied law and become an eminent advocate in Edinburgh, where he had befriended Enlightenment figures including architect Robert Adam, philosopher David Hume and economist Adam Smith. In 1760 William married the heiress Frances Pulteney, a cousin of the 1st Earl of Bath, with two major consequences: he changed his name from Johnstone to Pulteney; and he inherited fabulous wealth including large estates in Shropshire. He went on to invest in lands in North America, and in many developments in Britain including the famous Pulteney Bridge in neoclassical Bath, one of only four bridges in the world to have shops on both sides. And as if that wasn't enough to keep him occupied and amused, he entered politics and became an MP in seven successive parliaments, first representing Cromarty in Scotland, and then Shrewsbury, which he served for 30 years.

Recalling that Thomas Telford had previously given advice on repairs to the Johnstone mansion in Eskdale, and having decided to restore Shrewsbury Castle as a residence for

William Pulteney (1729–1805) by Thomas Gainsborough; and Pulteney Bridge across the River Avon in Bath, designed by Robert Adam and completed in 1774 (Wikipedia).

*Shrewsbury Castle (Geograph/David Smith); and Laura's
Tower, built by Telford (Geograph/Jeremy Bolwell).*

*Two of Shrewsbury's famous 17th-century timber
buildings: Council House Gateway; and Castle Gates
House (Geograph/Humphrey Bolton).*

himself, Pulteney contacted Thomas in 1787 and asked him to superintend alterations designed by Robert Adam. Living quarters were found in the castle for Thomas, and while there he found time to build a delightful tower in the grounds as a summer house for Pulteney's daughter Laura. Soon afterwards, no doubt thanks to Pulteney's influence, he was elevated to Surveyor of Public Works for the county of Shropshire. It marked the start of a fruitful collaboration and friendship – so much so that he became known in Shrewsbury as 'Young Pulteney'.

As county surveyor, Thomas had to familiarise himself with a wide variety of buildings and structures, some new, others needing inspection, repair or modification. Shrewsbury is famous for its ancient timber buildings, which must have delighted his eye and challenged any assumption that marble and stone were the only fine building materials. Shropshire would expand his horizons yet again – and increasingly as a designer in his own right rather than overseer of other men's plans. By the summer of 1788 he had more than ten jobs in hand including bridges, roads, drainage works, an infirmary, and a new county gaol.

The gaol project is particularly interesting, because it introduced Telford to one of the most remarkable men of the time, and awakened him to social issues that had so far escaped his concern. In a letter to his Langholm friend, Andrew Little, he reported:

> About ten days ago I had a visit from the celebrated John Howard, Esq. I say I, for he was on his tour of gaols and infirmaries; and those of Shrewsbury being both under my direction, this was, of course, the cause of my being thus distinguished … I accompanied him through the infirmary and the gaol. I showed him the plans of the proposed new buildings, and had much conversation with him on both subjects … You may easily conceive how I enjoyed the conversation of this truly good man, and how much I would strive to possess his good opinion. I regard him as the guardian angel of the miserable. He travels into all parts of Europe with the sole object of doing good, merely for its own sake, and not for the sake of men's praise.

John Howard was born in east London in 1726, the son of a successful upholsterer. On his father's death he inherited considerable wealth and settled on an estate in Bedfordshire. In 1773 he was appointed high sheriff of Bedfordshire and became responsible for its county gaol. He was shocked to discover that gaolers in English prisons were not salaried but lived off fees paid by prisoners for food, bedding, and other necessities. Penniless prisoners often lived in terrible conditions, and demands for payment before release meant that some stayed in gaol long after serving their sentences.

The entrance to Shrewsbury Gaol is little altered from Telford's original design. Above the doorway is a niche containing a bust of John Howard (Geograph/David Smith).

Howard's concerns reached Parliament and launched two 1774 reform acts. But fearing that they would be ignored, he embarked on tours of prisons in England, Scotland, and Ireland, followed by France, Holland, Flanders, several German states, and Switzerland. Over the next few years he added Denmark, Sweden, Spain, Portugal, and Russia to the list. At a time when travel was generally troublesome and frequently dangerous, he travelled nearly 50,000 miles at his own expense, making seven major journeys between 1775 and 1790. Tragically for a man of such profound humanitarian instincts, he contracted typhus while visiting military hospitals in Ukraine, and died there in January 1790. Seventy-six years later the Howard League for Penal Reform was founded in his honour – and continues its work today as the oldest prison reform charity in the UK.

So when John Howard visited Shrewsbury and met Thomas Telford, he had already devoted 15 years of his life to improving the treatment of prisoners at home and abroad. Not that Thomas's reverence for the hugely experienced campaigner spared him criticism: plans for the infirmary and gaol had to be substantially modified, the latter because Howard considered 'the interior courts too small, and not sufficiently ventilated'. But Thomas was happy to follow his guidance, not least because he realised that his new hero was physically and emotionally exhausted:

> he assures me that he hates travelling, and was born to be a domestic man. He never sees his country-house but he says within himself, "Oh! might I but rest here, and never more travel three miles from home; then should I be happy indeed!" But he has become so committed, and so pledged himself to his own conscience to carry out his great work, that he says he is doubtful whether he will ever be able to attain the desire of his heart – life at home. He never dines out, and scarcely takes time to dine at all: he says he is growing old, and has no time to lose. His manner is simplicity itself. Indeed I have never yet met so noble a being. He is going abroad again shortly on one of his long tours of mercy.

As Thomas may have guessed, and John Howard probably knew, that tour of mercy would be his last. He died at Kherson, on the shores of the Black Sea, less than two years after meeting his young admirer in Shrewsbury.

Another formative experience for Thomas Telford in 1788 was the collapse of St Chad's Church in Shrewsbury. Four centuries old, it had been in a lamentable state for some time, with a roof that directed rain onto the congregation. Thomas was called in, and after a rapid inspection of the structure announced to the churchwardens: 'Gentlemen, we'll consult together on the outside, if you please.' He was in no doubt that the building was in a dangerous condition and needed far more extensive repairs than a patched-up roof.

He now had a dispute on his hands. His damning report was rejected by the churchmen on the basis that 'professional men always wanted to carve out employment for themselves, and the whole of the necessary repairs could be done at a comparatively small expense'. A mason in the town was called in to cut away part of a stone pillar, and underpin it. Three days later, as the church bell struck four, its vibrations brought down the tower and demolished

Among the remains of Viroconium (Geograph/Jeff Buck).

the nave, forming, as Telford put it, 'a very remarkable ruin, which astonished and surprised the vestry, and roused them from their infatuation, though they have not yet recovered from the shock'. No doubt the affair did his reputation a power of good in the town; but a brand new St Chad's Church was quickly commissioned and built without any involvement by the county surveyor.

In the same year some extensive remains of a Roman city were unearthed on the outskirts of Wroxeter, a village about 5 miles from Shrewsbury. Viroconium, also known as Uriconium, is believed to have been the fourth-largest Roman settlement in Britain, with a population that reached 15,000 in its heyday. Samuel Smiles notes that:

> The situation of the place is extremely beautiful, the river Severn flowing along its western margin, and forming a barrier against what were once the hostile districts of West Britain. For many centuries the dead city had slept under the irregular mounds of earth which covered it, like those of Mossul and Nineveh. Farmers raised heavy crops of turnips and grain from the surface and they scarcely ever ploughed or harrowed the ground without turning up Roman coins or pieces of pottery … In fact, the place came to be regarded in the light of a quarry, rich in ready-worked materials for building purposes.

Curiosity was rekindled in 1788 when 'ready worked materials' were needed for a new blacksmith's workshop in the village. As building labourers dug down they disturbed some particularly interesting remains. Antiquarians arrived to inspect and advise, pronouncing the find to be part of a Roman bath house in a remarkable state of preservation; and William Pulteney, who counted the Lordship of the Manor among his many possessions, asked Thomas Telford to take charge of the necessary excavations. An extensive hypocaust (underfloor heating system) was uncovered, together with 'baths, sudatorium, dressing room, and a number of tile pillars – all forming parts of a Roman floor – sufficiently perfect to show the manner in which the building had been constructed and used'. Today the site is owned by English Heritage, and the remains of the bath house are open to the public, together with a visitor centre, shop, and museum.

As Thomas moved away from his original passion for stonemasonry towards wider responsibilities, he became increasingly involved in the problems of human relationships that beset all management. We have already seen that he had decided to tread carefully in Portsmouth to avoid conflict with the dockyard superintendent; to submit, albeit willingly, to design changes requested by John Howard; and to take on the stubborn churchmen of St Chad's. And now, for the first time, he found himself involved with a criminal fraternity outside, rather than inside, a county gaol. As Samuel Smiles notes:

> Among Telford's less agreeable duties about the same time was that of keeping the felons at work. He had to devise the ways and means of employing them without the risk of their escaping, which gave him much trouble and anxiety. "Really", he said, "my felons are a very troublesome family. I have had a great deal of plague from them, and I have not yet got things quite in the train that I could wish. I have had a dress made for them of white and brown cloth, in such a way that they are pye-bald. They have each a light chain about one leg. Their allowance in food is a penny loaf and a halfpenny worth of cheese for breakfast; a penny loaf, a quart of soup, and half a pound of meat for dinner; and a penny loaf and a halfpenny worth of cheese for supper; so that they have meat and clothes at all events. I employ them in removing earth, serving masons or bricklayers, or in any common labouring work on which they can be employed; during which time, of course, I have them strictly watched."

As an antidote we are offered some lighter fare – Thomas's tastes in theatre and music:

> Much more pleasant was his first sight of Mrs Jordan at the Shrewsbury theatre, where he seems to have been worked up to a pitch of rapturous enjoyment. She played for six nights there at the race time, during which there were various other entertainments … With the concert he was completely disappointed, and he then became convinced that he had no ear for music … "the melody of sound is thrown away upon me. One look, one word of Mrs Jordan, has more effect upon me than all the fiddlers in England … I felt no emotion whatever, excepting

*Famous fascinations
of Thomas Telford:
Dorothea Jordan
and Tom Paine
(Wikipedia).*

only a strong inclination to go to sleep … I suppose my ignorance of the subject,
and the want of musical experience in my youth, may be the cause of it."

So now we know: John Milton and *Paradise Lost*, yes; Shrewsbury theatre and especially
Mrs Jordan, certainly; but music of any kind, emphatically no – the rustle of wind in the trees
of Eskdale was more to his taste.

Mrs Jordan (1761–1816) was born in Ireland to an actress mother, and a stagehand
father who walked out when she was 13. Her impoverished mother saw thespian potential
in young Dorothea and put her on the stage. By the age of 21 she was appearing in towns
in Yorkshire and, 'having made her escape across the Irish Sea', started calling herself Mrs
Jordan. She was attracted to London's theatreland in 1785 and became known at Drury Lane
for a special talent in comedy and 'the most beautiful legs ever seen on the stage'. By the time
Thomas Telford saw her in Shrewsbury she was 27 years old and one of the most famous
actresses of the day.

What he and other besotted admirers may not have known was that Mrs Jordan had
indulged in five affairs with assorted lovers and produced several illegitimate children. They
could not have known that the pretty, witty, and intelligent actress would soon move upmarket
to become mistress of William, Duke of Clarence, later King William IV. The two lived together
more or less openly for 20 years and had at least ten children; and when separation occurred
in 1811 she was given an annual allowance and custody of their daughters (a rare concession
in those days) while he kept custody of the sons. Dorothea was forbidden to return to the
stage, but transgressed in 1814 to help pay off the debt of a son-in-law, angering the duke so
much that he took back the daughters and stopped her allowance. She fled to Paris to avoid
creditors and died in poverty a year later at the age of 55.

Another famous figure who fascinated Telford at about this time was Tom Paine (1737–
1809), English-American political activist, philosopher, and revolutionary. Born in Norfolk,

he emigrated to the British American colonies in 1774 and wrote the powerful, widely-read, pamphlet *Common Sense* to promote the cause of American independence. When the French Revolution broke out in 1789 he transferred his attention to France and produced his most famous polemic, *Rights of Man*, in support of the revolutionaries and against their appalled critics in the rest of Europe:

> In the spring of 1791 the first part of Paine's "Rights of Man" appeared, and Telford, like many others, read it, and was at once carried away by it. Only a short time before, he had admitted with truth that he knew nothing of politics; but no sooner had he read Paine than he felt completely enlightened. He now suddenly discovered how much reason he and everybody else in England had for being miserable ... Mr Paine had filled his imagination with the idea that England was nothing but a nation of bondmen and aristocrats. To his natural mind, the kingdom had appeared to be one in which a man had pretty fair play, could think and speak, and do the thing he would – tolerably happy, tolerably prosperous, and enjoying many blessings ... No one had hindered him; his personal liberty had never been interfered with; and he had freely employed his earnings as he thought proper. But now the whole thing appeared a delusion.

Unsurprisingly, it produced emotional turmoil in Telford. He had always accepted the social status quo, indeed had never given it much thought – in spite of being born in poverty and having had to haul himself up by his own bootstraps. But he was now the respected county surveyor of Shropshire, thanks, at least in part, to the patronage of William Pulteney, an extremely wealthy aristocrat.

To turn that on its head must have taken quite a bit of mental manoeuvring. Yet he seems to have managed it, at least temporarily. All of a sudden he had not the slightest difficulty offering revolutionary opinions about county magistrates, MPs, and the toffs who were 'carrying the country headlong to ruin'. With Paine's book as a guide, he felt competent to pronounce instant judgement on constitutional arrangements that had lasted for centuries. As he wrote to his Langholm friend Andrew Little:

> I am convinced that the situation of Great Britain is such that nothing short of some signal revolution can prevent her from sinking into bankruptcy, slavery, and insignificancy.

His dramatic change of heart was kept from the townspeople of Shrewsbury, the clergy of St Chad's, and 'the rosy-cheeked old country gentlemen' on whose goodwill, as county surveyor, he depended. But he had a near scrape with the man who had done most to help him since leaving Portsmouth – his patron, and increasingly his friend, William Pulteney.

As an MP, Pulteney enjoyed 'franking privilege' with the Post Office, allowing him to send personal mail free of charge. The system was widely abused, and it was generally expected that anyone with a parliamentary connection would get his friends' mail franked. But somehow Telford managed to use Pulteney's frank undetected, and posted a copy of Paine's 'Rights of Man' to Andrew Little in sleepy Langholm, where it created a sensation. Some of the inhabitants celebrated by drinking revolutionary toasts, and spent the next six weeks behind bars.

Pulteney was indignant when he discovered Telford's deception, but decided to forgive and forget. Telford, willing to forgive at least one of England's aristocrats, and increasingly aware that the liberty won by Parisians was turning to terror and bloodshed, became 'wonderfully reconciled' to the British system and the freedoms it gave him. From now on he would return to the things he knew best, leaving revolutionary politics to others.

At about this time he was asked to build a new bridge across the River Severn at Montford, 4 miles west of Shrewsbury. It was his first bridge in Shropshire, the first built anywhere under his superintendence, and surely what was needed to re-stabilise him after his obsessions with Dorothea Jordan and Tom Paine. Civil infrastructure executed 'in the service of man' would increasingly become Thomas Telford's hallmark, his life's calling, the practical expression of his special genius.

In the same year, 1792, that the Montford Bridge was completed, he was engaged to design and build a new parish church in Bridgnorth, Shropshire. His reputation was growing and, as mentioned earlier, he was increasingly accepted as an architect rather than a constructor of other men's designs. According to Samuel Smiles the church of St Mary Magdalen stands

> at the end of Castle Street, near to the old ruined fortress perched upon the bold red sandstone bluff on which the upper part of the town is built. The situation of the church is very fine, and an extensive view of the beautiful vale of the Severn is obtained from it. Telford's design is by no means striking, being, as he said "a regular Tuscan elevation; the inside is as regularly Ionic: its only merit is

Montford Bridge, built by Telford (Geograph/ Kevin Skidmore).

simplicity and uniformity; it is surmounted by a Doric tower, which contains the bells and a clock". A graceful Gothic church would have been more appropriate to the situation, and a much finer object in the landscape; but Gothic was not then in fashion – only a mongrel mixture of many styles, without regard to either purity or gracefulness. The church, however, proved comfortable and commodious, and these were doubtless the points to which the architect paid most attention.

Whether or not a 'graceful Gothic church' would have been more appropriate is, of course, a matter of taste. Smiles certainly knew how to damn with faint praise; but he was writing many years later, at a time when the Gothic Revival movement was in full flow in England, challenging neoclassical styles and harking back to the flamboyance of Christian medieval architecture. We should not be surprised that Telford's first adventure in church design drew heavily on his formative neoclassical experiences as a stonemason in Edinburgh and London. He was playing safe with what he knew, and many of us would agree with him; in any case the situation of his church is as fine today as ever, with extensive views over the Vale of Severn spread far below.

His next step was a tour of major towns and cities in the south of England to widen his knowledge of the 'best forms' of architecture. He visited Gloucester and Worcester, and was enchanted by a detour through the manufacturing districts of Gloucestershire, especially the fine scenery of the Vale of Stroud which seemed to him 'a smiling scene of prosperous industry and middle-class comfort'. He spent several days in Bath before taking the coach to London, where he re-examined the principal public buildings with a newly-expert eye; then returned to Shrewsbury via Oxford, which he adored, and Birmingham, which he loathed, judging it 'famous for its buttons and locks, its ignorance and barbarism'.

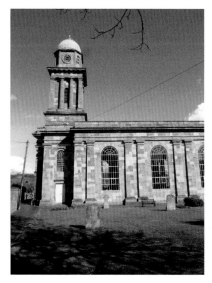 

St Mary Magdalen in Bridgnorth, designed by Telford and completed in 1795 (Geograph/James Allan; Wikipedia/ Andrew Abbott).

Looking over the River Severn from Bridgnorth High Town (Geograph/James Allan).

Back in Shropshire he resumed his work as county surveyor, overseeing roads and bridge repairs, and organising the convicts. The 35-year-old was still recognisably the affable Laughing Tam of his youth, but his independent spirit and increasing confidence meant he was no pushover. Above all, his hard work and competence continued to impress influential figures in the county and soon led to his next major, and entirely unexpected, move:

> An unforeseen circumstance, though not a hindrance, did very shortly occur, which launched Telford upon a new career, for which his unremitting study, as well as his carefully improved experience, eminently fitted him: we refer to his appointment as Engineer to the Ellesmere Canal Company.

Actually the appointment was not exactly as principal engineer to the company, but rather as resident engineer and assistant to one of the most experienced and respected canal engineers of the age, William Jessop. Nevertheless it was an extraordinary vote of confidence in a man who had absolutely no experience of canal engineering, offering him a ladder that would lead, step by step, to national and international fame.

The Ellesmere Canal was a huge undertaking planned for Shropshire, Cheshire, and Wales in the heady years of England's 'Canal Mania'. The original plan had envisaged a strategic north–south connection between Ellesmere Port near Liverpool, the manufacturing towns of the West Midlands, and Bristol, by cutting a canal from the River Mersey via Wrexham to the River Severn at Shrewsbury. Various branches would serve the needs of local landowners and industries. When shares were put on offer in the small town of Ellesmere in Shropshire in September 1792 there was a stampede to invest. A consenting Act of Parliament was passed in 1793, and William Jessop was appointed to take charge.

But the Ellesmere, as originally conceived, was never built. After endless commercial and political machinations by many interested parties the completed sections were eventually merged into the Shropshire Union canal system in 1846. In 1944 much of the 'Shroppie' was closed by the then owners, the LMS Railway; but they retained the canal's main line, and kept the branch to Llangollen open as a vital channel supplying water from the upper reaches of the River Dee. As a result today's 46-mile Llangollen Canal comprises two sections, originally part of the Ellesmere Canal and subsequently known as the Welsh and Llangollen branches of the Shropshire Union. Ironically, the rebranded canal carries more traffic than it did as a commercial waterway being hugely popular with leisure boaters and holidaymakers.

The challenges faced by Thomas Telford as he began to immerse himself in canal engineering are well worth our attention – not only for their own sake, but because they would stand him in great stead when he agreed, a few years later, to return to Scotland for one of his greatest projects – the Caledonian Canal.

One of the Llangollen's best-known features is the flight of six locks at Grindley Brook, about 8 miles from Hurleston Junction; on the sketch-map, these are the nearest to

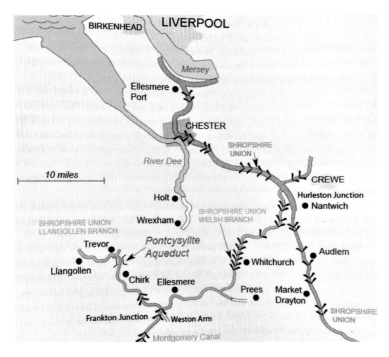

The Welsh and Llangollen branches of the Shropshire Union, now rebranded as the Llangollen Canal. Starting at Hurleston Junction on the main line of the Shropshire Union near Nantwich, it travels roughly southwest towards Ellesmere, then northwest to the Pontcysyllte Aqueduct and Llangollen.

Ascending the locks at Grindley Brook (Geograph/Roger Kidd).

Whitchurch. The first three come in close succession; but the next three are arranged as a staircase and require a certain amount of concentration. They presented Telford and Jessop with an interesting challenge.

For a boater, operating a staircase is very different from operating a normal flight, because a chamber can only be filled by emptying the one above, or emptied by filling the one below. This means that the whole staircase must be full of water (except for the bottom chamber) before a boat starts to ascend; or empty (except for the top chamber) before a boat starts to descend. Telford would find the experience of Grindley Brook a great help when, just a few years later, he tackled the stupendous flight of locks on the Caledonian Canal known as 'Neptune's Staircase'.

The charming bow-fronted lock keeper's cottage at the top of the Grindley Brook flight is a Telford design. If it looks more substantial than a cottage, this is because it also served as the toll house for collecting dues from passing narrowboats. It is easy to imagine the Ellesmere Canal Company applauding the new engineer on its staff who could not only tackle locks, but turn his hand to office buildings. And if the result lacks the neoclassical influences of Somerset House, or his church at Bridgnorth, it is because he decided to adopt an unfussy style more suited to state-of-the-art canal infrastructure, complete with angled windows that gave the lock keeper a good lookout over the comings and goings below. This experience, too, would be mirrored in buildings that Telford designed for the Caledonian Canal.

The lock keeper's cottage and toll house at the top of the Grindley Brook flight, designed by Telford (Geograph/ Jo Turner).

A few miles beyond Grindley Brook we come to the Ellesmere Tunnel, completed by Telford in 1802. Its 87 yards are modest by canal standards, and minimal compared to the famous 2,900-yard Harecastle Tunnel he would build on the Trent and Mersey Canal 25 years later; but tunnels, like locks, were strangers to him at the start of his canal career, and he was probably glad to be landed with a short one.

Soon after leaving the tunnel we reach Ellesmere, the small town after which the Ellesmere Canal, and its planned terminal at Ellesmere Port on the Mersey, were named. It was very much the centre of operations, with warehousing, cranage, dry dock, and workshops. Beech House, the company's impressive headquarters, was designed by Telford to provide a committee room, plans store, accounts office and apartments for resident engineer, agent, and accountant.

Beyond Ellesmere the canal strikes in a generally northwest direction towards Llangollen, via a one-mile embankment across flat and boggy Whixall Moss, two more tunnels, and two superb aqueducts.

Jessop and his new colleague knew they had some hard decisions to make about crossing the valley of the River Ceiriog and, even more dauntingly, the wider and deeper valley of the River Dee, far below the planned level of the canal. There was much to be said for preserving precious height by 'following the contour'; but surely no one would seriously suggest monumental bridge aqueducts of stone, topped by navigable troughs 'puddled' with clay in the traditional manner? Telford, after long discussions with Jessop, decided to break with accepted practice and recommend cast iron as a structural material in both aqueducts.

The first of these great structures is the 10-span, 70-foot-high, Chirk Aqueduct, completed in 1802. Telford specified iron plates for the base of its navigable channel, saving a lot of weight and permitting relatively slender arches. Chirk Aqueduct is followed almost at once by Chirk Tunnel, another major engineering work designed to preserve the canal's level. A silver ribbon of aqueduct water gives way to a short canal basin and then, almost immediately,

A cyclist passes over Chirk Aqueduct, enters Wales, and heads towards Chirk tunnel (Geograph/Roger Kidd).

to a leaden ribbon underground. Just over a quarter of a mile long, the tunnel was completed in November 1801. Telford considered the practice of 'legging' boats through canal tunnels degraded crews to mere beasts of burden and would have nothing to do with it. Instead he specified one of the first-ever tunnel towpaths for horses, supported on brick arches that allow water underneath and reduce the drag on passing boats.

The canal is now on course for the engineering structure that attracts thousands of boaters and sightseers to this corner of northeast Wales every year. Telford's wonderful Pontcysyllte Aqueduct above the River Dee is undoubtedly the high point of a UNESCO World Heritage Site – the 11-mile stretch of the Llangollen Canal from Chirk to the far side of Llangollen town. Rather than descend into the valley of the Dee with a long flight of locks, and climb up again on the far side, Telford daringly took to the air and designed a navigable iron trough 1,007 feet long, supported on 18 soaring masonry piers.

Pontcysyllte is arguably the greatest aqueduct built since Roman times, and one of the finest structural achievements of Britain's canal age. It represents Telford's determination to proclaim cast iron's virtues to the world: the ability to bear heavy structural loads, and a material that could do so with breathtaking flair and elegance. Ever since its completion in 1805 his 'stream in the sky' has been praised as an engineering masterpiece, a work of art that respects and even enhances its stunning natural environment. And it offers the extraordinary spectacle of narrowboats plying their trade 126 feet above the River Dee.

Trevor Canal Basin, at the far end of the aqueduct, was originally built as the next stage of the Ellesmere Canal, from where it would run north to serve the Wrexham coal and iron district before proceeding to Chester and Ellesmere Port on the Mersey. But the final stage of the main line was never built and the canal summit was left without its intended water supply. Instead a navigable feeder, daringly terraced along the valley side, was cut from Pontcysyllte to Llantysilio, 2 miles beyond Llangollen town, where it tapped water from the upper reaches

Pontcysyllte Aqueduct (Wikipedia/Akke Monasso).

Navigating Telford's 'stream in the sky' (Geograph/John M. Wheatley).

Horseshoe Falls at Llantysilio, built by Telford (Geograph/Ian Capper).

of the River Dee. Telford decided to build a 460-foot curved weir known as Horseshoe Falls across the river at this point, creating a large pond and guaranteeing a continuous feed to the canal. The weir and navigable feeder were authorised by an 1804 Act of Parliament, and completed in 1808.

By the time Thomas Telford had completed his work on the Ellesmere Canal he had accumulated valuable experience of locks, tunnels, cuttings, embankments, and aqueducts. Much of the design was his own, subject to the approval of William Jessop. Not surprisingly, he was increasingly regarded as an outstanding civil engineer who relished demanding challenges.

What is surprising – even dumbfounding – is that in 1801, even as the Ellesmere was entering the most critical period of its construction, Telford started to divert precious time and energy to an even greater canal project – the Caledonian Canal. It would bisect Scotland along the Great Glen from the Atlantic Ocean to the North Sea. He agreed to do a survey for the government, and in 1802 he reported on the project's feasibility. It was an opportunity close to his heart, one he could hardly refuse; so after two decades working in England and Wales, he returned to the land of his birth.

Scotland in Need

The internal communications of Scotland, which Telford did so much in the course of his life to improve, were, if possible, even worse than those of England about the middle of the 18th century. The land was more sterile, and the people were much poorer. Indeed, nothing could be more dreary than the aspect which Scotland then presented. Her fields lay untilled, her mines unexplored, and her fisheries uncultivated. The Scotch towns were for the most part collections of thatched mud cottages, giving scant shelter to a miserable population ... The common people were badly fed and wretchedly clothed, those in the country for the most part living in huts with their cattle.

These comments on the state of Scotland at the time of Thomas Telford's birth may strike you as exaggerated and ungracious. But their author, Samuel Smiles, had been born and brought up in Scotland and had studied medicine at Edinburgh University. Far be it from me to accuse him of anti-Scottish sentiment.

There is little doubt that Tom's entry into the world took place in a troubled land. It was only 50 years since the Act of Union had joined the parliaments of England and Scotland to form Great Britain, and in the meantime the Highlanders had launched two Jacobite rebellions that threatened what, to most of the English, had seemed an established Protestant order. The second rebellion, led by Bonnie Prince Charlie in 1745, raised an army large enough to take Edinburgh before marching as far south as Derby – a profoundly unnerving crisis for the London government. Although subsequent retribution, swift and brutal, was principally aimed at the Highlanders and their clan system, a continuing air of crisis and uncertainty in Scotland reduced the chances that her citizens would feel much improvement in their lives, at least in the short term.

Samuel Smiles also describes a veritable slave class in Scotland

who were bought and sold with the estates to which they belonged, as forming part of the stock. When they ran away, they were advertised for as negroes were in the American States … Money was then so scarce that Adam Smith says it was not uncommon for workmen, in certain parts of Scotland, to carry nails instead of pence to the baker's or the alehouse. A middle class could scarcely as yet be said to exist, or any condition between the starving cottiers and the impoverished proprietors, whose available means were principally expended in hard drinking … and were, for the most part, too proud and too ignorant to interest themselves in the improvement of their estates.

In any case there was a longstanding problem with Scottish infrastructure away from the cities and large towns which, insofar as it existed at all, was in a lamentable state. England had about five times the population of Scotland but nearly 40 times the wealth. Not surprisingly, most Scottish roads were even worse than those in England. Bridges capable of taking wheeled vehicles were few and far between. Communications, always difficult, were often impossible in winter. The remarkable improvements in roads, bridges, harbours, buildings, and agriculture that took place in Telford's lifetime became major drivers of economic and social progress, finally raising country districts from their slough of despond. But in 1757, the year of his mother's bereavement, that was all for the future – a future to which her infant son would make an outstanding contribution.

Fortunately by the time Thomas entered the labour market in the 1770s conditions in the Scottish border country, including Eskdale, were beginning to change for the better. The reivers had largely given up their raids in exchange for more cooperative activities, and the young Duke of Buccleuch was busy improving his estates. As we have seen, Thomas became an apprentice stonemason, then worked for the duke before moving to Edinburgh. By 1781 he was in London, propelling himself along a career path that would lead via Portsmouth and Shrewsbury to canal building in Shropshire and the hill country of northeast Wales.

By the turn of the century the English canal network had been growing for about 40 years, with a huge spurt during the 'Canal Mania' of 1792–3. About 50 new waterways had been opened in England, soon to be joined by Jessop's and Telford's partly-completed Ellesmere Canal. There were also four in Scotland that were well known to Telford.

The first and by far the most important of the Scottish quartet was the Forth & Clyde Canal, designed by the English civil engineer John Smeaton, begun in 1768, but completed only in 1790, after a series of funding crises. Originally intended as a route for seagoing vessels between the Firth of Forth in the east and the Firth of Clyde in the west, it passed through the narrow neck of the Scottish Lowlands with an important basin at Port Dundas in Glasgow. The 35-mile canal included 39 locks, about 70 feet long by 20 feet wide – dimensions that seemed generous at the time, but far less so as the shipbuilding industry developed. Fortunately the canal was also a valuable route for horse-drawn barge traffic, especially after a new waterway, the Union Canal, was cut in 1822 from the Forth & Clyde, near its eastern end, to the city of Edinburgh.

From the 1860s onwards the fortunes of the Forth & Clyde were buttressed by a remarkable series of sturdy coasters known as Clyde Puffers. Built about 66 feet long to fit inside the locks, the steam-powered, steel-hulled puffers could carry up to 100 tons of cargo. About 400 were launched over a period of 70 years, in three classes: inland or 'inside'; 'estuary'; and seagoing or 'outside'. Most were built in yards along the canal. Many 'outside' puffers traded throughout the Hebrides and became much-loved by the isolated island communities they served.

The closest I ever got to a Clyde Puffer – although I failed to realise it at the time – was on holiday in Scotland in 2006. On the way from Fort William towards Glenfinnan and Mallaig, my wife and I decided to picnic on the quayside at Corpach Basin, the southwestern terminal of the Caledonian Canal. And there, almost beside us, was a doughty little steamer of dubious date but undoubted charm. I have since realised that she was a puffer, and probably the one that played the role of *Vital Spark* in a famous BBC television series 'The Tales of Para Handy', set in the 1930s, broadcast between 1959 and 1995 and, at the time of writing, still available. Covering the crazy adventures of a puffer's crew around the coasts and islands of western Scotland, I suspect it gives a far more romantic view of puffer life than the harsh realities faced by real-life crews in cramped accommodation, often tossed on boisterous seas.

The Forth & Clyde Canal was closed in 1963 and became semi-derelict. But not forgotten. It has since been restored to something like its former glory, and reconnected to the Union Canal by a superb piece of modern canal engineering known as the Falkirk Wheel, the only rotating boatlift in the world. Telford, man of iron and water, would have loved to see it.

The second Scottish canal well known to Telford was the 9-mile Crinan in Argyll and Bute, started by John Rennie (a fellow Scot) in 1794. It has 15 locks and 6 swing bridges, and connects Lochgilphead on a small inlet of Loch Fyne with Crinan on the Sound of Jura. This offers a navigable shortcut between Glasgow, Oban and the Hebridean islands, without the need for a long diversion around the exposed Mull of Kintyre. Like the Forth & Clyde,

Vital Spark, one of the last Clyde Puffers, became a TV star (Wikipedia/Jmb).

The Forth & Clyde restored: one of its 39 locks (Wikipedia/CVPont), and the Maryhill Road Aqueduct (Geograph/Thomas Nugent).

The Falkirk Wheel reconnected the Forth & Clyde Canal to the Union Canal in 2002, linking Glasgow to Edinburgh (Wikipedia/Sean MacClean).

the Crinan became strongly associated with the Clyde Puffers that did so much to supply Highland coastal communities.

However the Crinan's construction and initial use were plagued with financial and engineering problems. Originally opened in 1801, it suffered a major failure of the canal bank in 1805, and a burst summit reservoir in 1811 which ruined some of the locks. The canal company, headed by the Duke of Argyll, requested help from the government, and Telford was called in to assess the problems. He recommended modifying the canal's course, improving the locks, and replacing timber swing bridges with cast iron.

On the Crinan Canal (Geograph/M.J. Richardson).

A holiday visit by Queen Victoria in 1847 made the Crinan a tourist attraction, and a few years later steamer companies based in Glasgow started offering passages to Oban via horse-drawn boats along the canal. Like the rest of Britain's waterways, the Crinan went into severe decline in the 20th century; but it is now fully restored and functional, a popular route for about 2,000 leisure craft annually between the Firth of Clyde and the west coast of Scotland. A regular visitor is Clyde Puffer VIC32 which, at the time of writing, offers 'Para Handy' holidays at a very gentle pace to steam enthusiasts.

The other two Scottish canals known to Telford were far more modest affairs. The Stevenston (1792), lockless and a mere 2 miles long, used barges with a capacity of about 15 tons to carry coal from local pits to a harbour on the coast of Ayrshire about 25 miles southwest of Glasgow. The Monkland Canal (1794), 12 miles long with 18 locks, was cut for barges 71 feet long by 14 feet wide which transported coal from the mining area east of Glasgow into the city. It was subsequently linked to the Forth & Clyde and became very profitable. But these two artificial waterways can have had little impact on Telford as he contemplated the huge challenge of a Caledonian Canal. His major point of reference in Scotland was undoubtedly the Forth & Clyde.

This seems a good moment to step back a little and consider a wider question: why was a radical improvement in transport infrastructure desperately needed in 18th-century Britain?

Canals and roads were certainly topics for intense discussion by the middle of the century, but it was mostly talk, not action, and the shocking state of the roads meant that packhorses were still the main means of bulk transport overland. Together with the main rivers, new canals and roads were urgently needed as arteries of Britain's burgeoning industrial revolution. It was in England that the need was felt most urgently; and in England where it first began to be remedied – by Francis Egerton, Duke of Bridgewater.

The duke, exasperated by the slow and unreliable packhorse trains that carried coal 8 miles from his mines in Worsley to the expanding town of Manchester, decided to commission a canal in 1760. The flat terrain needed no locks, but the River Irwell was in the way and had to be crossed. The duke was well aware of navigable aqueducts on the Canal du Midi in France, but the very idea of a bridge carrying a canal over a river seemed ridiculous to many of his English contemporaries. Yet the duke pressed ahead with his Bridgewater Canal and appointed 44-year-old James Brindley as engineer.

Born in 1716 into a family of yeoman farmers in the Peak District of Derbyshire, Brindley had been educated at home by his mother before becoming a millwright, wheelwright, and eventually a talented repairer and even designer of new types of machinery. His first task for the Duke of Bridgewater was to design and build a bridge aqueduct over the River Irwell near the village of Barton. The design he came up with was about 600 feet long and 36 feet wide, an enormous structure for its time. It was completed in just 11 months; but on the first day water was admitted it began to sag. Poor James Brindley, overcome with anxiety, retired to his bed at the local tavern and left a colleague to address the problem with a partial rebuild. The aqueduct was re-puddled with clay and allowed to stand until the following spring, by which time the mortar had set and the danger of collapse had passed.

Crowds came to view a wonder of the age, amazed to see a single horse hauling linked barges across Brindley's aqueduct while a team of horses struggled to haul a single barge against the flow of the Irwell 40 feet below. Just as importantly for the Duke of Bridgewater and his customers, the new-fangled 'navigation' cut transport costs to such an extent that the price of coal in Manchester fell by a half. No doubt the Scottish pit owners of Stevenston and Monklands were well aware of the extraordinary profits being generated south of the border.

James Brindley's growing fame put him in great demand and he was soon appointed engineer to the 93-mile Trent & Mersey Canal linking the River Trent near Nottingham to the

Long and narrow: Brindley's Lock no.41 on the Trent and Mersey Canal (Geograph/Star Watcher).

Bridgewater Canal and the River Mersey – a huge project to which he would devote six years of unremitting attention before his untimely death in 1772. The canal had a big influence on British civil engineers and, as we shall see, Telford became personally involved in its upkeep.

Locks were new to Brindley, so he decided to experiment with a model. The design he came up with was unusually narrow, able to take a barge up to 70 feet long but only 7 feet wide – dimensions that, according to many canal historians, cast a long shadow over the developing canal network. We have inherited many 'narrow canals' that use Brindley's lock dimensions, plus a smaller number of 'wide canals' with locks of similar length but about twice the width. It is worth noting that the Forth & Clyde Canal, opened 13 years after the Trent & Mersey, has lock dimensions 70 feet by 20 feet, of standard length but unusually wide for its time.

Brindley was also new to tunnels. Of the four needed, the Harecastle in Staffordshire, one of the longest canal tunnels ever built, was by far the most problematic. Beset by unexpected changes of rock type, flooding, and lack of ventilation, its seven-year construction outlasted Brindley by five years. He gave it no towpath, expecting boatmen to leg their way through by lying on the top of the boat and pushing on the slimy sides or roof of the tunnel with their feet. They emerged from the far end completely exhausted and 'as wet from perspiration as if they had been dragged through the canal itself'. The 2,880-yard passage took up to three hours, while the lucky horses were escorted over Harecastle Hill.

Growing traffic and the painfully slow process of legging soon made the Harecastle Tunnel a serious bottleneck. Boats would pile up at one end, waiting for others to complete their passage in the opposite direction, and there were many furious battles between claimants

The northern entrances to Thomas Telford's Harecastle Tunnel of 1827 and, in the background, James Brindley's tunnel of 1777, now abandoned (Wikipedia/Akke Monasso).

for the 'first turn through'. So in 1824 it was decided to build a wider parallel tunnel. Thanks to advances in civil engineering and the talents of its designer – none other than Thomas Telford, by then aged 67 – it was completed in three years. We have already seen that Telford, in his work on the Ellesmere Canal 25 years earlier, refused to subject boat crews to legging and gave his tunnels a towpath. And so it was with the Harecastle.

The two tunnels were restricted to one-way traffic and used in conjunction, but following a partial collapse of Brindley's tunnel in 1914 Telford's took over completely. Nowadays narrowboats complete the passage in about 40 minutes, their crews protected from diesel fumes by powerful forced ventilation.

In the last few years of his life Brindley planned, and partly completed, more than 300 miles of canals. His early successes proved the viability of horse-drawn canal boats, and industrialists in other parts of the country were soon clamouring for a piece of the action. Initially the traffic was mainly inland; but as the network reached coastal port cities, more cargo was exchanged with seagoing ships for import and export. This trend was reflected in Scotland by the Forth & Clyde Canal, which allowed trading along the canal and gave access to the North Sea at one end and the Irish Sea at the other. And in due course it would be strengthened by Telford's Caledonian Canal which, in effect, joined the North Sea and the Baltic Sea to the Atlantic Ocean.

Judged by today's standards, we may wonder why the ability to horse-haul a barge carrying 30 tons of coal, sand, china clay, or manufactured goods at average speeds of about 2 miles per hour was considered such a huge improvement on existing modes of transport. Surely wheeled vehicles on improved roads would be a faster and better solution? And where new roads were needed, surely they could be built more cheaply than canals by avoiding the huge expense of locks, embankments, cuttings, tunnels, and the occasional aqueduct?

Not necessarily – and our Victorian author, Samuel Smiles, has plenty to say about it. The third volume of his famous series 'Lives of the Engineers', published in 1862, was updated two years later with an 'Account of Early Roads and Modes of Travelling', including a chapter on Scotland. His style propels us back to Victorian times in ways that a modern writer would struggle to achieve – slightly quaint, but thoroughly engaging, and redolent of a very different age:

> Freedom itself cannot exist without free communication – every limitation of movement on the part of the members of society amounting to a positive abridgment of their personal liberty. Hence roads, canals, and railways, by providing the greatest possible facilities for locomotion and information, are essential for the freedom of all classes, of the poorest as well as the richest.

In addition to packhorses, simple unsprung wagons hauled by teams of horses had been used for centuries to transport goods and, to a lesser extent, people; but passengers found that mode of travel

> so tedious, by reason they must take waggon very early and come very late to their innes, that none but women and people of inferior condition travel in this

sort ... The waggons made only from ten to fifteen miles in a long summer's day; that is, supposing them not to have broken down by pitching over the boulders laid along the road, or stuck fast in a quagmire, when they had to wait for the arrival of the next team of horses to help drag them out.

If the situation in England was bad, in Scotland it was lamentable:

The misery of the country was enormously aggravated by the wretched state of the roads. There were, indeed, scarcely any made roads throughout the country. Hence the communication between one town and another was always difficult, especially in winter ... Single-horse traffickers, called cadgers, plied between the country towns and the villages, supplying the inhabitants with salt, fish, earthenware, and articles of clothing, which they carried in sacks or creels hung across their horses' backs. Even the trade between Edinburgh and Glasgow was carried on in the same primitive way, the principal route being along the high grounds west of Boroughstoness, near which the remains of the old pack-horse road are still to be seen.

Travel conditions for people were hardly better than for goods. It was not until 1749 that the first public conveyance between Scotland's two major cities, the 'Glasgow and Edinburgh Caravan', started to run – and it took two days to complete the distance. As late as 1763 there was only one stagecoach between Scotland and London; it set out from Edinburgh once a month and took anything between 10 and 15 days, depending on the weather. Sensible passengers made their wills before departure. The coach was no faster in 1781 when Thomas Telford ignored its doubtful benefits and rode to London on Sir James Johnstone's horse.

Rural districts were even more isolated. The West Highlands were virtually impenetrable. More surprisingly, there were no proper roads in the southern counties of Dumfries and Galloway, and even aristocrats had difficulty moving around:

When the Marquis [sic] of Downshire attempted to make a journey through Galloway in his coach, about the year 1760, a party of labourers with tools attended him, to lift the vehicle out of the ruts and put on the wheels when it got dismounted. Even with this assistance, however, his Lordship occasionally stuck fast ...

Unfortunately durable and affordable roadmaking, especially in difficult country, was little understood. Telford himself would become a major innovator and driving force, but in the meantime human travel in most of Scotland was on foot for the poor, on horseback for the lucky few. Bulk transport for goods and manufactured products remained with the packhorses and the wagons.

Why were the new industrialists, including coal owners, becoming so fed up with the service packhorses provided? The main reason is also the obvious one: the small weight of goods a packhorse could carry – about 240 pounds, distributed evenly between two side-panniers strapped across its back. True, packhorses could be arranged in long trains, typically 12 to 20 and occasionally up to 40, and coaxed or bullied as much as 25 miles a day along tracks that were impossible for wheeled vehicles. But compare this with the load hauled by a single horse coupled to a 72×7-foot barge that fitted snugly into one of Brindley's new-fangled canal locks: about 30 tons or 67,000 pounds, equivalent to 280 packhorse loads – and transportable by a crew of two at roughly the same average speed.

Packhorses had other disadvantages. The small stocky animals, often named Galloways after the Scottish county where they were first bred, needed a lot of feeding and human attention; many were lamed by their arduous duties; and a train was often impeded by dreadful conditions underfoot or the difficulty of passing others travelling in the opposite direction. The animals often wore a hoop or collar of bells to warn of the convoy's approach:

> in many parts of the path there was not room for two loaded horses to pass each other, and quarrels and fights between the drivers of the packhorse trains were frequent as to which of the meeting convoys was to pass down into the dirt and allow the other to pass along the bridleway.

According to Smiles conditions in the 1760s were hardly better than 100 years before and led to innumerable quarrels. Injuries were common, deaths occasional. Some packhorse men, 'rude Russian-like rake-shames', simply got rid of approaching horseback riders with booze-fuelled threats of violence.

Much of Britain's early industrialisation was concentrated in the north of England. By the mid-18th century huge numbers of packhorse trains were crossing the Pennine Hills loaded with coal, limestone, salt, fleeces, and cloth. Packhorse bridges were built over streams and small rivers. The importance of major routes was reflected in popular jingles and rhymes that served as aides-memoire for the packhorse men, and in the growth of packhorse inns, many of which have survived into our own times.

A long packhorse train (Wikipedia).

The packhorse bridge at Carrbridge (Wikipedia/Bert Kaufmann).

The packhorse bridge near John O'Groats (Geograph/Stanley Howe).

Scotland also had – and still has – its packhorse bridges. Probably the most famous survivor can be found in the village of Carrbridge, close to the Cairngorms National Park. Built in 1717, it was severely damaged in the 'Muckle Spate' (great flood) of 1829; today it is classed as a scheduled ancient monument. Another survivor, less well known but better preserved, was built in 1651 near John O'Groats. Those of us who live south of the border may be surprised to learn that packhorses were as familiar to country people in the far northeast of Scotland as they were in the far southwest of England.

Part 2

Return to Scotland

The Caledonian Canal

Telford returned to Scotland in 1801 to survey the route for a canal through the Great Glen between Fort William and Inverness. By this time the condition of his native land was clearly on the mend – especially around Edinburgh and Glasgow, up the east coast to Aberdeen, and on to Inverness. The Scottish Enlightenment of the 18th and early 19th centuries, a movement which asserted the importance of human reason over traditional authority, was championing practical benefits for society including advances in political economy, architecture, and engineering. Telford had already seen and felt some of its effects during his stay in Edinburgh 20 years earlier. As he travelled north from Shropshire towards the Great Glen, he crossed the Forth & Clyde Canal, clear proof that Scotland was enjoying infrastructure improvements near its two great cities. But the West Highlands were, as yet, virtually untouched – and waiting for him.

A Caledonian canal had long been a topic for conversation in Scotland, and the great James Watt had done a survey 28 years before Telford arrived on the scene. As Samuel Smiles remarks:

Thomas Telford in middle age, a portrait by Scottish painter Henry Raeburn (ICE).

The formation of a navigable highway through the chain of locks lying in the Great Glen of the Highlands, and extending diagonally across Scotland from the Atlantic to the North Sea, had long been regarded as a work of national importance. As early as 1773, James Watt, then following the business of a land surveyor at Glasgow, made a survey of the country at the instance of the Commissioners of Forfeited Estates. He pronounced the canal practicable, and pointed out how it could best be constructed. There was certainly no want of water, for Watt was repeatedly drenched with rain while he was making his survey, and he had difficulty in preserving even his journal book. "On my way home", he says, "I passed through the wildest country I ever saw, and over the worst conducted roads."

In view of what we have learned about roads in 18th-century Scotland, it is hardly surprising that James Watt found them 'the worst conducted' he had ever seen. Perhaps we should be more surprised that this remarkable man, associated the world over with early steam engines, had started his professional life as a land surveyor; and that the Forfeited Estates mentioned by Samuel Smiles were but one example of the retribution enacted by London on Scotland following the Jacobite rebellions of 1715 and 1745, which ordered the seizure of assets owned by Highlanders judged guilty of high treason.

Government indecision ensured that James Watt's proposal, which included 32 locks with a capacity about seven times greater than Brindley's aptly-named 'narrow' variety, was never actioned. Nor was a subsequent one, prepared in 1793 by John Rennie. But by the end of the century, with Britain at war with France and Napoleon on the rampage, inland ship canals were being actively considered to protect the Royal Navy's 32-gun frigates from French privateers operating along British coastlines.

There were further pressing reasons for a Caledonian canal: vessels sailing from the northeastern ports of England and Scotland for Liverpool, Ireland or America had to confront the notorious Pentland Firth between mainland Scotland and the Orkney Islands, often against gale force winds and vicious tidal streams, or go even further north to pass between Orkney and Shetland. Not only were both these routes dangerous; they could be extremely time-wasting. As an example: two vessels were despatched on the same day from Newcastle-upon-Tyne, one bound for Liverpool by the north of Scotland, the other for Bombay (Mumbai) by the English Channel and the Cape of Good Hope. And yes, you have probably guessed: the latter reached its destination first. Another example: an Inverness vessel, sailing on Christmas Day for Liverpool, reached Stromness Harbour in Orkney on 1 January. Three months later it was still there – stormbound.

The canal was also seen as part of a wider infrastructure initiative to improve trade and communications across the Highlands and, equally important for the government, to curb emigration by providing desperately needed employment. Telford's time had come.

When asked to survey the route in 1801 Telford wrote to Watt, whose firm Boulton & Watt had recently supplied a 30-inch beam engine for pumping duties on the Ellesmere Canal:

> I have so long accustomed myself to look with a degree of reverence at your work, that I am particularly anxious to learn what occurred to you in this business while the whole was fresh in your mind … If I can accomplish this, I shall have done my duty; and if the project is not executed now, some future period will see it done, and I shall have the satisfaction of having followed you and promoted its success.

He was tempted to follow the line recommended by Watt, but only after making a thorough survey of his own. A key requirement was an adequate supply of water at the summit level – the *sine qua non* for a successful canal. He and Jessop were already facing serious water supply problems on the Ellesmere, so the last thing he wanted was similar trouble on the Caledonian, especially as he intended recommending a canal with huge, thirsty locks.

The natural rift known as the Great Glen that cuts diagonally through the Highlands from Fort William to Inverness is 60 miles long. About 38 miles are covered by three freshwater lochs, and the highest of these, Loch Oich, was the natural choice for the canal's summit level and main water supply. But Telford realised it was small and shallow, and would need a back-up source.

Fortunately Loch Oich is fed by Loch Garry; and Loch Garry is fed, in turn, by Loch Quoich, some 20 miles west of the Great Glen in beautiful but desolate mountain country. Even today Loch Quoich is considered remote; back in 1801 it was virtually unvisited, and unknown to outsiders. Typical of a man who left no stone unturned, Telford decided to seek it out for himself.

While researching material for this book, I have often been reminded of our own travels over the years in the West Highlands. When my wife and I first journeyed from the Great Glen at Invergarry to Loch Garry, and on via Tomdoun to Loch Quoich, we never imagined we were retracing the steps of Thomas Telford. Nor did we know that he had descended to

Loch Quoich (Geograph/Richard Webb).

Loch Hourn, a majestic sea loch on the west coast, bypassed Munros that we had climbed in earlier days, and returned to the Great Glen by a more northerly route:

> from Loch Hourn I travelled by track scarcely less rugged to the top of Glen Elg and over the steep mountain of Raatachan to the top of Loch Duich; from thence I travelled along the vestiges of a Military Road up Glen Shiel, down a part of Glen Moriston and over a rugged mountain to Fort Augustus.

How he completed the best part of 100 miles through rough country remains a mystery to me: as there were no proper roads, apart from 'the vestiges' of a military one, presumably he would have ridden, rather than walked. But on his own, or in company? And where did he sleep? Anyway he returned to the Great Glen reassured that Lochs Quoich, Garry, and Oich could provide his canal with a reliable water supply. And as we shall see, he had traversed rough tracks that he would later turn into desperately needed Highland roads.

In the following year, 1802, Telford continued his survey, this time accompanied by William Jessop. Meanwhile a Royal Navy captain based at Fort Augustus had taken soundings in Loch Oich and the two other large freshwater lochs that formed part of the intended line, Loch Lochy and Loch Ness. Telford now produced plans for a canal 100 feet wide at the surface and 50 at the

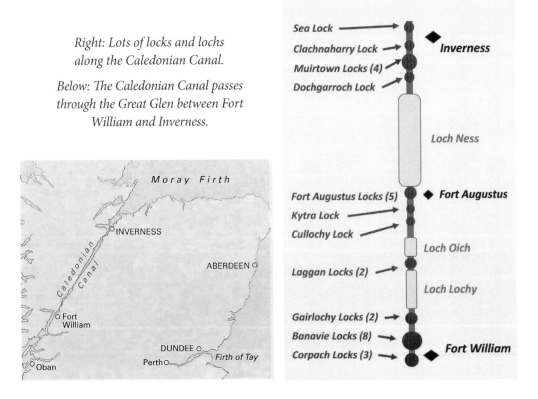

Right: Lots of locks and lochs along the Caledonian Canal.

Below: The Caledonian Canal passes through the Great Glen between Fort William and Inverness.

bottom, with 29 locks each 170 feet long by 40 feet wide – with a capacity about 40 times greater than the standard 'narrow' model specified for the Ellesmere. His recommendations were accepted, the Caledonian Canal Commission was set up, and consenting Acts of Parliament were passed in 1803–4. For the second time in a burgeoning career, Telford was appointed principal engineer to one of the most demanding canal projects ever undertaken, with Jessop as consultant.

The canal, as built, is illustrated in the accompanying diagram. Vessels coming from the Atlantic approach Fort William via two sea lochs, and enter the canal by a tidal lock and basin at Corpach near Fort William. Next comes a famous flight of eight locks at Banavie known as 'Neptune's Staircase', followed by two more locks at Gairlochy. Loch Lochy is the first freshwater loch, separated by the two Laggan Locks from Loch Oich, the summit of the canal. From there it is 'downhill all the way' via 14 locks to the Moray Firth and the North Sea.

It is time to start our own journey along the Caledonian, beginning in the same place as the workforce did in 1804 – Corpach, which Samuel Smiles describes as situated

> amidst some of the grandest scenery of the Highlands. Across the Loch is the little town of Fort William, one of the forts established at the end of the seventeenth century to keep the wild Highlanders in subjection. Above it rise hills over hills, of all forms and sizes, and of all hues, from grass-green below to heather-brown and purple above, capped with heights of weather-beaten grey; while towering over all stands the rugged mass of Ben Nevis – a mountain almost unsurpassed for picturesque grandeur.

A visit to Corpach, with Ben Nevis in the background (Paul A. Lynn).

Personally I doubt whether the view of Ben Nevis from Corpach is unsurpassed for picturesque grandeur, and can only assume that Smiles had never ventured west to Loch Quoich and Loch Hourn!

The works at Corpach gave Thomas Telford and his newly-appointed (but unrelated) resident engineer John Telford a great deal of trouble. For a start they had to build some very basic houses for the masons and a barrack for the labourers, followed with almost indecent haste by a brewery to tempt the men away from the 'pernicious habit of drinking whisky'. Even so there were labour troubles including angry disagreements over pay rates and late arrivals of the necessary cash from Inverness. It took time for the situation to settle down.

There were also engineering problems. The chamber of the great sea lock that isolates Corpach Basin from the tidal rise and fall on Loch Linnhe and Loch Eil had to be excavated out of solid rock. The necessary coffer-dam could only be kept clear of water by the ceaseless efforts of a 20 horsepower Boulton & Watt steam engine, sent in parts from Birmingham by canal and sea. It arrived along Loch Linnhe aboard the 44-foot sloop *Corpach,* one of two vessels ordered by the commissioners. Her duties would include the import of granite, rubble stone, and limestone; Welsh oak for sea-lock gates; and iron for railways, locks, and swing bridges, delivered from 'Merlin' Hazeldine's foundry at Plas Kynaston and 'Iron Mad' Wilkinson's ironworks at Bersham in northeast Wales. Meanwhile temperamental Highland workers hacked and blasted their way through rock, praying that Watt's new-fangled engine would not let them down.

We leave Corpach Basin and start along an attractive mile of level cutting, accompanied by a walker's dream, the Great Glen Way. And then comes the best-known feature of the entire canal, summarised by Samuel Smiles:

> It was necessary to climb up the side of the hill by a flight of eight gigantic locks, clustered together, and which Telford named Neptune's Staircase. The ground passed over was in some places very difficult, requiring large masses of

Corpach Basin: in the background, the sea lock onto Loch Eil; in the foreground, the lock leading from the basin towards Neptune's Staircase (Geograph/Chris Heaton).

The Great Glen Way begins alongside the Caledonian Canal near Corpach (Geograph/Jim Barton).

Approaching Neptune's Staircase (Geograph/S.J. Dowden).

Swing bridge open: the railway line from Fort William to Mallaig (Geograph/Clive Perrin).

Ascending the staircase (Geograph/Bill Henderson).

No shortage of water (Geograph/Martin Horsfall).

Left: Lock no.5, with Telford's house in the background (Geograph/Ian Taylor).

Upper right: Banavie Wharf at the top of the staircase (Geograph/Bill Kasman).

Right: Kayaks at Banavie (Geograph/Bill Kasman).

embankment, the slips of which in the course of the work frequently occasioned serious embarrassment.

Two bridges guard access to the staircase: a swing bridge carrying the famous railway between Fort William and Mallaig, a line beloved by connoisseurs of the world's great rail journeys as well as fans of the Harry Potter films; and a rival road bridge carrying the A830 along a parallel route (pioneered by Telford himself as far as Arisaig, and later extended to Mallaig).

The staircase raises the canal through 60 feet. It is a 'real' one (rather than an 'apparent' one with large side ponds), so the upper gates of one lock serve as the lower gates of the next. Boats climbing the staircase are committed to passing through all eight locks before others can descend. If you are a boater you may be excused for finding the whole thing a bit daunting, especially if you are used to the English canals pioneered by James Brindley. Neptune's locks were the largest ever constructed in Britain.

Originally 36 capstans had to be rotated in strict order to control the locking sequence, and many keepers were needed to operate the system smoothly and safely. It took about half a day to clamber up the staircase, but 20th-century mechanisation has cut the passage time to around an hour and a half. Of course, you may have to wait for another hour and a half before starting the ascent!

This huge civil engineering work employed up to 60 quarrymen and 120 masons, and took four years to complete. Telford, a highly experienced master mason, claimed that 'there does not seem to be the slightest imperfection in this immense mass of building'. He could not have known that disastrous storms and flooding in 1834 (the year of his death) would cause collapses here and elsewhere on the canal.

As we reach Lock no.5 an attractive listed building comes into view: a house of a size and quality far above most canal architecture, set in its own grounds, with a three-bay window. A glance back at the photo of Telford's lock keeper's cottage at the top of the Grindley Brook staircase on the Llangollen Canal is enough to convince us that this is also a Telford design, further proof of his determination to give the new canals their own domestic architecture – unfussy, elegant, functional.

Banavie village, at the top of the staircase, offers guesthouses, hotels, and a long pontoon for visiting boats. The Great Glen Way passes through the village, then mostly follows the canal towpath for 8 miles to Gairlochy, where a pair of locks raises the water level another 16 feet to meet beautiful Loch Lochy. On the way it passes the Loy Sluices, which ditch excess canal water into the River Lochy – and, unnoticed by most boaters, crosses three aqueducts which on most waterways would be considered important structures. We should remember that the canal intersects multiple mountain streams or 'burns', some welcome, others superfluous, and nearly all capable of changing rapidly from a trickle to a torrent. Water management and control has always been a big issue on the Caledonian.

Loch Lochy is 9 miles long and plunges to a maximum depth of 531 feet (162 metres), making it the third deepest freshwater loch in Scotland after Loch Morar (1,017 feet/310 metres) and Loch Ness (755 feet/230 metres). An infamous battle between Clan Macdonald and Clan

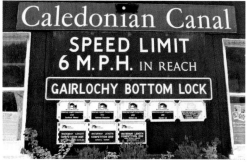

Upper left: Clyde Puffer VIC32 cruises past the Loy Sluices between Banavie and Gairlochy (Geograph/Jim Barton).

Left: Dutch barge Ros Crana occupies most of Gairlochy Top Lock's length (Geograph/Jim Barton).

Above: A good rule and some competition trophies: a noticeboard at Gairlochy Bottom Lock (Geograph/Jim Barton).

Fraser was fought at the northeastern end of Loch Lochy, near Laggan, in July 1544. It became known as Blar-ne-leine, or Field of the Shirts, allegedly because the heat of the day caused the Highlanders to cast off their heavy tartan plaids. Tragically, the beauty of the place did nothing to soften the violence: about 800 men failed to return to their families that evening.

The two Laggan Locks at the far end of Loch Lochy complete our ascent to the summit. And now for a surprise. The short stretch of canal from here to Loch Oich includes one of the greatest feats of civil engineering ever undertaken in Britain at the time – the Laggan Cutting, 40 feet deep and 2 miles long. Author Tom Rolt, himself an engineer, had no doubt of its significance and the reason for its comparative lack of fame:

> By their very remoteness, Telford's great works in the Highlands never excited the public wonder and acclaim which works of far smaller magnitude aroused in populous England. Thus the mighty summit cutting at Laggan by which Telford drove the Caledonian Canal from Loch Oich to Loch Lochy dwarfs in its scale almost all the works of the later railway builders which the world was soon to marvel at. Railways and barrow runs were both used in the work and as soon as a part of the excavation was deep enough to be flooded the Donkin

Heading out into Loch Lochy (Geograph/Bill Kasman).

dredging machines were brought in to increase the depth. But few visitors saw the Laggan cutting while it was being driven, and when it was finished the water which filled it and the trees which were planted to bind its sides soon concealed the scale of it.

The same holds true today: the Caledonian Canal's enormous locks are plain for all to see; the width of its stretch between Banavie and Loch Lochy is clearly impressive; but the mighty Laggan Cutting, like Telford's southern approach embankment to the Pontcysyllte Aqueduct, conceals its monumental scale with an arboreal cloak and can easily go unheeded by the boater or towpath walker.

Loch Oich, far shallower than Loch Lochy, also needed dredging. In 1816 it played host to one of the steam-powered bucket dredgers mentioned by Tom Rolt, constructed at the Butterley Ironworks in Derbyshire by a famous and versatile engineer, Bryan Donkin. It had arrived in parts and was assembled on the shores of the loch, where it made an extraordinary spectacle.

The 4 miles of Loch Oich, ending at Cullochy Lock, were soon followed by Kytra Lock and a short length of canal to Fort Augustus, where a staircase of locks leading down to Loch Ness gave Telford a particular headache. Tom Rolt describes it well:

Left: Entering a Laggan Lock (Geograph/Jim Barton).

Right: A walk through the Laggan cutting (Geograph/Jim Barton).

Left: From Cullochy towards Loch Oich (Geograph/Jim Barton).

Right: The five-lock staircase at Fort Augustus,
leading down to Loch Ness (Geograph/Anne Burgess).

The greatest difficulty was encountered in building the bottom lock chamber at Fort Augustus. This had to be sunk far below the level of Loch Ness and a bed of loose, permeable gravel was encountered through which the waters of the loch poured in almost uncontrollable quantity. Here the last and largest of the three Boulton & Watt engines which had been supplied to the Commissioners was pressed into service. This was of 36 h.p., having a cylinder 4 feet in diameter and a piston stroke of 8 feet – hardly a handy unit to haul about the Highlands. Yet even this titan, labouring night and day, could not cope with the constant influx and finally both the Clachnaharry [at the northern end of the canal] and Corpach engines as well had to be pressed into service before the waters were mastered and the chamber built.

Passing down the Fort Augustus staircase we might spare a thought for three of the greatest steam engines ever built at that time, and the engineers who installed and nurtured

Looking along Loch Ness from Fort Augustus (Geograph/Jim Barton).

Urquhart Castle beside Loch Ness (Wikipedia/Sam Fentress).

them, little imagining that their work would give pleasure to holidaymakers two centuries later.

The lockless 23-mile cruise along Loch Ness should be calm and relaxing – provided there is not a strong headwind from the northeast, or unwelcome attention from a cryptozoological monster. Urquhart Castle, on a headland beside the A82 near Drumnadrochit, is certainly worth a visit. Dating from the 13th century, it played a role in the Wars of Scottish Independence in the 14th century, was largely abandoned by the middle of the 17th, and partially destroyed in 1692 to prevent a takeover by Jacobite rebels. Today it is in state hands and classed as an ancient monument.

The deep waters of Loch Ness gave Telford no trouble, but little Loch Dochfour at the far end needed dredging. After this the canal follows the course of the River Ness for a few miles towards Inverness, then branches off towards Clachnaharry and the Beauly Firth, past a marina, four Muirtown Locks, and a mooring basin.

Clachnaharry: the quirky name suggests we are about to meet something extraordinary, and it turns out to be the largest sea lock of its time, sunk into a morass of mud where the canal meets the sea, an achievement considered by many professional engineers to rival and even exceed Neptune's Staircase and the great Laggan Cutting. Tom Rolt describes the challenge:

> The shore of the Beauly Firth at Clachnaharry shelves very gradually. This meant that the entrance lock for the new canal must be carried out 400 yards beyond the shore line in order that ships could enter it at any state of the tide. On the site chosen for this lock an iron bar was driven into the mud; it sank fifty-five feet before it found solid bottom. The consistency of this mud was such that oak piles rebounded from it as though they were on springs at each blow of the driver and all hope of constructing a coffer-dam had to be abandoned. The solution adopted was to build out from the shore a great clay embankment. On the site of the lock this was weighted with stone and left for six months during which time it slowly settled, extruding the soft mud from beneath it. When this process of consolidation was complete, piles were driven and great oblong timber frames secured to them. These formed the main members of a huge coffer-dam within which the artificial earthwork was excavated to the full depth of the lock chamber. Horse-driven chain pumps and a 9 h.p. Boulton & Watt beam engine worked constantly to keep the lock pit clear of water. In the bottom of the excavation a bed of rubble stone bound in water-lime mortar two feet thick at the centre and five feet thick at the sides was laid down as a foundation for the lock chamber invert and the side walls which were six feet thick. It was a work of the greatest difficulty which was not completed until 1812, but it proved perfectly satisfactory.

So ends our cruise along the Caledonian Canal – a good moment to reconsider the roles of Jessop, Telford, and the resident engineers who took responsibility for one of the greatest artificial waterways ever built in Europe.

Left: Caley Marina near Muirtown Locks (Geograph/Craig Wallace).

*Right: The great clay embankment and sea lock stretching out into the
Beauly Firth at Clachnaharry (Geograph/Richard Murray).*

*Left: The Clachnaharry embankment contains the last section of the canal and acts as a basin
for vessels waiting to pass through the sea lock into the Beauly Firth (Geograph/Craig Wallace).*

*Right: A narrowboat leaves Clachnaharry sea
lock for the Beauly Firth (Geograph/Martin Clark).*

William Jessop, consulting engineer, was nearly 60 when construction work began in
1804, and he died ten years before the first vessel passed along the canal from ocean to ocean
in 1822. As on the Ellesmere Canal, he was a reassuring figurehead, an engineer of enormous
experience, final arbiter of important engineering decisions.

Thomas Telford, principal engineer, could hardly oversee all the day-to-day construction
and management himself, but relied on a team of resident and assistant engineers attracted
from far and wide to the Great Glen. An important figure was John Telford, whom we have
already met struggling with labour disputes and the great sea lock at Corpach. Another was

Matthew Davidson, who had been working on the Ellesmere Canal and was enticed away from Wales by Thomas Telford, for whom he had the greatest respect.

We know little about John Telford, except that he had previously worked as a toll collector on the Ellesmere Canal at Chester. He must have made a good impression because in 1804 he was promoted to resident engineer for the 8-mile stretch of Caledonian Canal between Corpach and Loch Lochy including, crucially, Neptune's Staircase. By April 1805 he was directing a workforce of about 400 men, mostly locals, and probably wondering what on earth he had let himself in for. He wrote anxiously to his boss, Thomas Telford:

> Last Saturday was pay day and a very disagreeable one it was; notwithstanding the men was all informed when you was here that those upon days wages would only receive 1/6d per day, they refused to take it. Nor do I suggest it will be settled without going before a Justice. Mr Wilson and myself were in eminent [*sic*] danger of our lives; yet notwithstanding we would not give way to one of them, tho they threatened much and were on the point of using violence many times.

Tragically he died just three years later, at the age of 36, well before Neptune's Staircase was completed. It seems that irksome management responsibilities and the world's greatest lock staircase had proved overwhelming. Was it a rare example of Thomas Telford, widely admired for his ability to select, promote, and support able colleagues, choosing the wrong man? Perhaps he had been drawn to unrelated John Telford by the coincidence of both surname and Christian name shared with the 'unblameable' shepherd of Eskdale, the father he had never known. In any case Alexander Easton was quickly appointed successor for the Corpach to Loch Lochy section, at the surprisingly young age of 20. He lasted the course, remained in post until the canal was opened, and died in 1854 at the age of 67.

The other key figure in Telford's engineering team, Matthew Davidson, was a man admired and trusted for his hard work, loyalty, and expert knowledge of stonemasonry. Originally attracted by Telford to the Ellesmere Canal, he had been entrusted with raising the soaring stone piers of the Pontcysyllte Aqueduct. He now swapped northeast Wales for the Highlands, and was given responsibility for about 500 men and the 8-mile stretch of canal between Loch Ness and Clachnaharry including, crucially, the great embankment and sea lock leading out into the Beauly Firth.

Older than John Telford and much tougher, Matthew Davidson was by all accounts an extraordinary character. Largely self-educated, he was a prodigious reader who attracted the nickname Walking Library. He held strong political opinions, vehemently expressed, and loathed republicanism and unbridled power in any form. A fitness fanatic, he advocated bathing in icy water, including that of the Beauly Firth, as a cure for every ailment; and as a Lowland Scot he maintained an ill-concealed distaste for the Highlands and their clans. But his more humane side had grown to love the people of Wales, the Welsh girl he had married, and the three sons she had borne him before he left for the Highlands. He brought so many masons with him that his house at Clachnaharry was viewed by locals as a Welsh colony. His youngest son James took over after his untimely death in 1819. Father, as so often, must have been a hard act to follow.

Having concentrated initially on the two end sections of the canal and their great sea locks, Thomas Telford moved most of his huge workforce to the 11-mile central section between Loch Lochy and Fort Augustus. By 1819 the works were nearing completion and he hoped to see the canal open for business in 1820 or early 1821. But the huge difficulty posed by the bottom lock at Fort Augustus thwarted his ambition, and it was October 1822 before the first boat sailed through the Great Glen from sea to sea – 11 years behind schedule and three times over budget. Even so there were great celebrations, reported by the local paper, the *Inverness Courier*:

> The doubters, the grumblers, the prophets and the sneerers, were all put to silence, or to shame; for the 24th of October was at length to witness the Western joined to the Eastern sea. Amid the hearty cheers of the crowd of Spectators assembled to witness the embarkation, and a salute from all the guns that could be mustered, the Voyagers departed from the Muirtown Locks at 11 o'clock on Wednesday with fine weather and in high spirits. In their progress through this beautiful Navigation they were joined from time to time by the Proprietors on both sides of the lakes; and as the neighbouring hamlets poured forth their inhabitants, at every inlet and promontory, tributary groups from the glens and the braes were stationed to behold the welcome pageant, and add their lively cheers to the thunder of the guns and the music of the Inverness-shire militia band, which accompanied the expedition.

Yet it must be admitted that the canal's critics had plenty of ammunition. The channel through the summit cutting and Loch Oich was just 12 feet deep, not the promised 20, in spite of huge efforts by the steam dredgers. Engineering defects, including a slippage in the Laggan Cutting and further trouble with the Fort Augustus locks, persisted. To make matters worse, the commercial prospects were slipping away like water through a leaky lock gate: the size of shipping was increasing, outgrowing the canal's restricted draught; steamships were appearing over the horizon, robbing the sea routes around the north of Scotland of their historic terrors; and the Royal Navy's case for the canal had dwindled with the defeat of Napoleon at Waterloo. French warships and privateers would no longer threaten British frigates.

Sadly, after 18 years of construction and a series of huge engineering challenges bravely faced, Telford was subjected to criticism by Parliament and public. This seems unfair; commercial and strategic issues were not his responsibility: he had been tasked by government to design and construct a canal, not gaze into a crystal ball. A week may be a long time in politics, but not in a massive civil engineering project. If he was guilty of anything, it was his optimistic estimate of the cost and time needed for completion.

Unfortunately those persistent defects, mainly on the central section of the canal, meant that by the 1840s it was in such a sorry state that Parliament debated major restoration or even abandonment. Fortunately the restorers prevailed, and engineering works in 1843–7 were supervised by Telford's successors, James Walker and Alfred Burges. The canal was repaired, improved, and deepened, attracting vessels engaged in the Baltic trade and making significant contributions to the Highland economy through fishing and tourism. Passenger steamers became popular following a visit by Queen Victoria in September 1873. Service in World War I for the transport of mines and munitions gave it a temporary shot in the arm. But in spite of occasional lifelines, the waterway continued its inexorable fall from grace well into the 20th century, slowly wasting away, like the rest of the British canal system, as railways and roads came to dominate the transport infrastructure.

In June 1996 I noticed an article about the canal in *The Times* newspaper, cut it out, and slipped it into my copy of Tom Rolt's biography of Thomas Telford. Headed 'Caledonian Canal facing closure', the article is still there – and paints a decidedly gloomy picture:

> The Caledonian Canal, one of the greatest engineering works of early 19th century Britain and a key contributor to the Highlands economy, is in danger of closure unless £20 million is raised for emergency repairs. The canal is leaking so badly that most of the original 29 locks designed by the engineer Thomas Telford 200 years ago are in need of extensive maintenance … Jim Stirling of British Waterways said yesterday that without it the canal would certainly close, possibly as soon as next summer.

Luckily the clarion call was heeded: next summer came and went, and many more have followed. Further restoration has been carried out on major structures, including Neptune's Staircase, the tourists' favourite. As a result the wonderful scenery and variety of experience the canal gives to boaters, walkers, and sightseers is still on offer, as is the excitement, for children of all ages, of sailing along Loch Ness with eyes open for an improbable monster.

Recent restoration work on Neptune's Staircase reveals its extraordinary scale (Geograph/ Alan Reid).

Today the Caledonian Canal is internationally recognised as an engineering landmark of the early 19th century. A memorial plaque at Fort Augustus reads:

INTERNATIONAL HISTORIC CIVIL ENGINEERING LANDMARK

CALEDONIAN CANAL

CONSTRUCTED 1804–22 (DREDGING CONTINUING), REFURBISHED 1843–47

ENGINEER: THOMAS TELFORD (1757–1834)

CONSULTING ENGINEER: WILLIAM JESSOP (1745–1814)

CONTRACTORS: JOHN SIMPSON, JOHN WILSON & JOHN CARGILL

THIS 60 MILE LONG 110 FT WIDE SHIP CANAL ACROSS SCOTLAND BETWEEN THE NORTH AND IRISH SEAS, CONSTRUCTED USING STATE-OF-THE-ART TECHNOLOGY ON AN UNPRECEDENTED SCALE, WITH 28 LOCKS 170 FT OR 180 FT LONG, 40 FT WIDE, AND 25 FT DEEP, THEN A SERIES OF THE LARGEST EVER BUILT. THE CANAL SIGNIFICANTLY ADVANCED HIGHLAND DEVELOPMENT AND ENGINEERING KNOWLEDGE.

PRESENTED TO BRITISH WATERWAYS IN THE 250TH ANNIVERSARY YEAR OF TELFORD'S BIRTH

BY

THE INSTITUTION OF CIVIL ENGINEERS

AND

THE AMERICAN SOCIETY OF CIVIL ENGINEERS

4 JULY 2007

Actually, I have not quite finished with the Caledonian Canal – and for what you may consider an unlikely reason. Almost exactly 200 years ago it was visited by two men, one an eminent engineer and lover of poetry, the other a renowned poet who longed to see and record the engineer's work in the Highlands of Scotland. We are about to follow a journey they made together in 1819, which included time spent along the length of Scotland's most spectacular canal.

On tour with a poet

Rendezvous in Edinburgh

On 17 August 1819 two famous men met in Edinburgh at the start of an ambitious tour. Robert Southey, age 45, held the prestigious title of poet laureate; Thomas Telford, 17 years his senior, was engaged on a wide variety of Highland infrastructure projects including the Caledonian Canal, roads, bridges, and harbours. Both were at the height of their careers.

ROBERT SOUTHEY
From the portrait by T. Phillips, R.A.

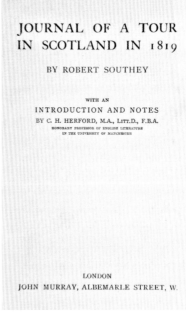

JOURNAL OF A TOUR
IN SCOTLAND IN 1819

BY ROBERT SOUTHEY

WITH AN
INTRODUCTION AND NOTES
BY C. H. HERFORD, M.A., LITT.D., F.B.A.
HONORARY PROFESSOR OF ENGLISH LITERATURE
IN THE UNIVERSITY OF MANCHESTER

LONDON
JOHN MURRAY, ALBEMARLE STREET, W.

The frontispiece and title page to Robert Southey's book (photos: Paul A. Lynn).

Among my book collection is a first edition copy of Robert Southey's *Journal of a Tour in Scotland in 1819,* an account of what might reasonably be described as a working holiday with family attachments. The idea seems to have come from Southey's friend John Rickman, who wished to bring along his wife, two children, and a certain Miss Emma Piggot, described by Telford as 'a young lady of prepossessing appearance and agreeable manners'. For Telford it was a chance to inspect work in progress throughout the Highlands, enlivened by mixed company. For Southey it promised dramatic mountain scenery for comparison with his home territory – the English Lake District – and an insight into the professional activities of an engineer he had long admired.

Fortunately Southey and Telford, who were about to spend six weeks in extremely close company, hit it off straight away:

> Mr Telford arrived in the afternoon from Glasgow, so the whole party were now collected. There is so much intelligence in his countenance, so much frankness, kindness and hilarity about him, flowing from the never-failing well-spring of a happy nature, that I was upon cordial terms with him in five minutes.

Robert Southey (1774–1843) had had a start in life about as different from Thomas Telford's as is possible to imagine. Born in Wine Street, Bristol, he was educated at Westminster School and Balliol College, Oxford. He later said of his university education that 'all I learnt was a little swimming ... and a little boating'. By the time he was 20 his first collection of poems had been published and he was busy experimenting in a writing partnership with fellow poet Samuel Taylor Coleridge. Like many young idealists, and especially those of an artistic temperament, he had become obsessed by the French Revolution and its dream of a new social order (something Telford had flirted with during his time in Shrewsbury). Together with Coleridge and several others Southey dreamed of creating an idealistic community on the banks of the Susquehanna River in the northeastern United States, based on simple wants, shared possessions, a love of literature and science, and the promise that each young man could 'take to himself a mild and lovely woman for his wife'. But Southey was the first to go cold on the plan, and when they failed to agree it was abandoned.

Instead Southey married Edith Fricker at St Mary Redcliffe in Bristol in 1795. She was a sister of Coleridge's wife, Sara. The newlyweds made their home at Greta Hall, Keswick, in the Lake District, where they managed on his modest income. Also living there (after Coleridge abandoned them) were Sara and her three children, and the widow and son of another poet, Robert Lovell. Southey became known as an English poet of the Romantic School, one of the Lake Poets, together with Coleridge and Wordsworth.

In addition to his poetry, Southey was a prolific letter writer, literary scholar, essayist, historian, traveller, and biographer. His biographies include the life and works of John Bunyan, Oliver Cromwell, and Horatio Nelson. As a man of many interests he was willing and able to take a wide view of the artistic and intellectual climate of the time, including civil engineering projects widely considered 'art' as well as 'science'.

Originally a radical supporter of the French Revolution, Southey became disillusioned by its violence and turned increasingly towards a conservative view of politics and society. As his fame grew he was accepted by the Tory establishment, and was appointed poet laureate by King George III in 1813. By the time he met Telford in Edinburgh six years later he had become something of an establishment figure who could 'look forward with complacent security to the renown awaiting him in the next and later generations'. But he retained an inquisitive mind and an interest in the living conditions of his fellow citizens, both rich and poor.

Southey was attacked by less successful poets who kept the radical faith; but he protected himself well enough from most of them, and their continuing undercurrent of criticism hardly affected his eager anticipation of the forthcoming Highland tour. Little did he imagine that the king would die a few months after the tour ended, and that his Funeral Ode would be seen by critics as a further proof of sycophancy. Lord Byron wrote a brilliant parody of the ode, an onslaught from which Southey's reputation as a poet would never fully recover. Today he is probably better remembered for his children's 'Story of the Three Bears', the original Goldilocks story, than for his adult output as a Lake Poet.

John Rickman (1771–1840), the third male member of the touring party, was the son of a clergyman. Born in Northumberland, educated at Guildford Grammar School and Oxford University, he was 48 years old when he and his family joined Southey and Telford in Edinburgh. Rickman served as secretary to two parliamentary commissions established in 1803: one to oversee the building of roads, bridges, and harbours in Scotland; the other to keep an eye on the construction of the Caledonian Canal. Telford was appointed a commissioner on both, so the two men had worked closely for many years and become friends. Rickman was an expert

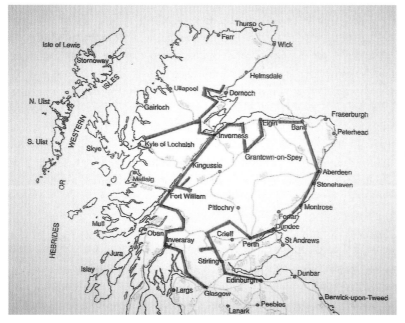

The route followed by Southey, Telford and Rickman in 1819.

statistician, best remembered nationally as the creator of the United Kingdom census. He drafted the 1800 Act 'For taking an account of the population of Great Britain, and of the increase or diminution thereof', and was instrumental in organising the first four censuses. He was elected to the Royal Society in 1815 and, although not an MP, became a well-respected public servant of the House of Commons. His friendship with Telford was to prove lifelong.

Telford, Southey, Rickman, and entourage spent three days in Edinburgh, sightseeing and visiting friends. The capital city must have stirred many memories in Telford, who had arrived there nearly 40 years earlier to immerse himself in the art of stonemasonry and the study of architecture. He was now to act as the party's guide on a tour that would illustrate just how far he had advanced professionally since those early days. The accompanying map shows their route: first inland from Edinburgh, then broadly anticlockwise in a great sweep up the east coast via Aberdeen to Inverness, along the Caledonian Canal (then under construction) from Inverness to Fort William, and on to Glasgow via Inveraray. Several diversions along the route helped satisfy Telford's professional commitments, Southey's love of mountain landscapes, and Rickman's natural curiosity.

Southey's account of their travels certainly shows an enquiring mind. His book of nearly 300 pages goes far beyond what might be expected of a poet laureate and gives us valuable and often amusing glimpses into the history, culture, landscapes, farming, and social conditions of Scotland, as witnessed by an inquisitive Englishman experiencing the Highlands for the first time. It also paints a picture of Telford far removed from his reputation as an engineer obsessed by his work, and illuminates his social skills with a huge variety of people, from the highest to the humblest.

To Stirling, the Trossachs, and Perth

Southey and company left Edinburgh in the early morning of 20 August:

> Rose at five, packed my trunk, inserted in my journal all the remaining mem-
> oranda concerning Edinburgh, and now the Coach is at the door. These horses
> are to take us to Linlithgow, whither those which are engaged for the whole jour-
> ney were sent forward yesterday … At a turnpike not far from Edinburgh is this
> inscription, "Whisky, porter, and ale: uppiting for horses", which comical word
> seems to mean up-putting … A cartload of harvesters past us on their way afield.
> Men and women in abundance were busy at this chearful work; but they seemed
> less active and less regular in their movements than English labourers would have
> been. No pastures here, and few hedges – hence an open and somewhat of a for-
> eign appearance in the country. 16 miles to Linlithgow.

Southey found Linlithgow 'decayed, dirty, and dolorous' – apart from its majestic palace, which they visited: 'We were shown in it the apartments wherein Mary and Charles the First were born. The quadrangle is fine: one side appears to be of the early Tudor age; one is in the viler stile of James the First, where the windows are made in imitation of woodwork'. The next town, Falkirk, was more prosperous because of a recent connection to the Forth & Clyde

Linlithgow Palace (Geograph/Bill Kasman).

The statue of Robert the Bruce at Bannockburn (Geograph/Stanley Howe).

Canal, which they passed under 'by an arch so dangerously low, that it might easily prove fatal to a traveller on the outside of a stage coach'.

The road to Stirling took them past the site of the Battle of Bannockburn, 'the only great battle that ever was lost by the English – their only disgraceful defeat. At Hastings there was no disgrace. Here it was the army of Lions commanded by a stag'. On the way they watered the horses and took the opportunity to refresh themselves 'with luxurious Twopenny, which is bottled small beer, as weak as Mr Locke's metaphysics, as frothy as Counsellor Phillips's eloquence and, when the cork has been drawn a few minutes, as vapid as an old number of the Edinburgh Review'. Not, it seems, an attractive swig.

Close to Stirling they entered on a short stretch of new road and inspected a bridge with a huge circle over a single arch – 'the first bridge which I have seen in this form: the appearance is singular and striking'. It was one of Telford's.

Actually, Stirling's most historic bridge – then and now – had been built about 300 years before their arrival; this was a timber one, made famous by William Wallace when he used it to help defeat Edward I's forces at the Battle of Stirling Bridge in 1297. It was replaced in the 1400/1500s by the stone Stirling Old Bridge, which played a part in the Jacobite rising of 1745, when an arch was removed to hinder Bonnie Prince Charlie and his forces as they marched south. It remains one of the finest medieval stone bridges in Scotland.

The city of Stirling, once the capital of Scotland, is visually dominated by its splendid palace-cum-castle. The tourists could hardly pass it by, and it made a big impression on the English poet. The external part, built by James V, intrigued him for its 'architectural caprice', heavily ornamented with grotesque monsters and the kings and queens of Scotland. The view

Stirling Old Bridge (Geograph/J. Thomas).

Stirling Castle: (Geograph/M.J. Richardson).

from its 'commanding eminence' over the richly cultivated valley of the River Forth appealed even more than that from Edinburgh Castle, and Southey rested briefly on a seat intended for 'the aged and infirm, who had long resorted to the spot, on account of its warmth and shelter from every wind'. On obtaining entry they found the palace converted into apartments for army officers, its chapel into an armoury, and the pulpit abandoned in a corner: 'A more dignified use, more becoming the nation and the monarchy, might be devised for these ancient and memorable buildings.'

Stirling, Gateway to the Highlands, is a 'huge brooch that clasps Highlands and Lowlands together'. Historically it had strategic importance as the final bridge over the widening River Forth as it heads towards the Firth of Forth and the open sea. 'He who holds Stirling holds Scotland', a well-known dictum, emphasised that travel between Edinburgh and the north almost inevitably meant crossing Stirling Bridge. No doubt Telford pondered all this as they headed out of the city towards Callander, 16 miles distant.

Southey now relates the first incident relating to wheeled vehicles on Scottish roads:

> one of the near horses stumbled, owing I believe to a loose stone in the road, and fell: it was on a slight descent, and he was dragged a few yards before the leaders could stop the carriage: the postillion fell under him, and cried out piteously "O my leg, my leg!". Happening to be on the box with Willy, I saw this frightful

accident. By good fortune it was close to a blacksmith's, and he being not afraid of horses was of more service than all the other persons who presently gathered round. The driver was soon extricated; and when the poor fellow found that he could both stand and walk, he said in a cheerful tone that he had had a broken leg before … so that half an hour's delay seemed to be all the harm.

They must have been relieved to reach Callander at nine o'clock that evening and find accommodation 'in the fine large inn of this poor place'. An hour later they were sitting down to hot and cold salmon, lamb chops, and good bottled porter, followed by a night's rest 'in an apartment not less than twenty four feet square, and twelve in height – for what can it be wanted in such a situation? On the chimney piece are a black cock and hen stuffed. We have none of these birds in Cumberland.' It is unclear how many of the party of seven were expected to share the space with Scottish birds.

Next morning they set off in search of Highland lochs, peaks, and heather. The area they were about to enter, the Trossachs, was something of a diversion from their main route up the east coast and it seems likely that Telford, aware that his poet companion would relish an early opportunity to compare spectacular landscapes with his beloved Lake District, had recommended the northerly detour. In addition he and Southey were well aware that the Trossachs, and one of its major scenic attractions, Loch Katrine, were intimately associated with Walter Scott, whose 1810 narrative poem 'The Lady of the Lake' had caused a national sensation.

The 10 miles from Callander to Loch Katrine were 'not only very bad, but in no slight degree dangerous, for in many places the slightest deviation from the track would either upset the carriage by the rocks on one side, or precipitate it down a crumbling bank on the other'. Arriving at the end of the loch they marvelled at the steep slopes of Ben Venue, which reminded Southey of Helvellyn as seen from Ullswater, and discovered a small inn that had been a farmhouse before Scott catapulted the Trossachs into fashion. Carriages stopped there and guides were in readiness – signs of a tourist boom which continues to this day, encouraged by pleasure craft that ply their trade along the loch. Southey recalls:

> We embarked as soon as we came to the waterside, and it was amusing to hear the boatmen relate with equal gravity the exploits of Bruce, Cromwell, and Rob Roy, and the incidents of Scott's poem as connected with the scenery of the Lake. The Island would be more beautiful if it were not so lofty; and it is too much covered with wood. We landed not far from it, on the right shore (that is, looking up the Lake), and ascended to a point whence the whole may be seen … the end is certainly unsurpassed in its kind – perhaps unequalled by anything I have ever seen.

So far so good: Southey the poet loved Loch Katrine, and glossed over any misgivings about Scott's huge popularity and his refusal to accept the laureateship six years earlier because it denoted sycophancy towards the Crown. Southey, having picked up the 'poisoned chalice' himself, seems keen to show that some trace of his youthful radicalism has survived:

Cruisers alongside the jetty at the end of Loch Katrine (Geograph/M.J. Richardson).

Last year the Duke of Montrose sold the woods on Ben Venue, which was then compleatly clothed with fine trees, for the paltry price of 200£. It seems incredible that for such a sum he should have incurred the obloquy and the disgrace of disfiguring, as far as it was in his power to disfigure, the most beautiful spot in the whole island of Great Britain. Notwithstanding the remoteness of the situation the purchasers have made 3000£ by their bargain. The scenery must have suffered much, but not so much as might be supposed … But this did not enter into his Grace's calculations; he is fairly entitled to all the vituperation which is bestowed upon him by the visitors to Loch Katren.

Whether Southey expected his words to reach aristocratic eyes and ears is unclear. The journal would not be published in his lifetime – the manuscript was in family hands until 1864, remained in private hands until 1894, and then found a home at the Institution of Civil Engineers, where it was exhibited during centenary celebrations in 1928. At this point it was offered to publisher John Murray, who produced 500 copies (one of which is now mine) the following year.

Before leaving Loch Katrine I should like to indulge in a little engineering history. Water from the loch has been delivered by aqueduct to the city of Glasgow since 1859, and although the scheme has been augmented and improved over the years most of the original structures remain in place. It reminds me of Telford's pioneering work on the Ellesmere Canal in northeast Wales half a century earlier, and especially his Pontcysyllte Aqueduct.

Glasgow's water had been drawn from wells and streams until 1807, when Thomas Telford and James Watt, who were already corresponding about the Caledonian Canal, built a new pumping station at Dalmarnock on the River Clyde, a mile or two upstream of central Glasgow. Further works were added as the city expanded, but the river became increasingly polluted and a stinking danger to public health. A severe outbreak of cholera in 1848–9 claimed over 4,000 lives. Glaswegians desperately needed the crystal waters of a Highland loch.

In 1852 John Frederick la Trobe Bateman, a civil engineer already famous for designing Manchester's water supply system, was brought in to review the options. He reported that raising the level of Loch Katrine by 4 feet with a small dam could provide the city with 50 million gallons of water a day, supplied by a 34-mile aqueduct with a reservoir near the Glasgow end. He advised that 'no other source' would do – a view supported by fellow engineers Robert Stephenson and Isambard Kingdom Brunel. The scheme was approved by Glasgow Corporation, and construction began in 1855.

The original aqueduct network was remarkable, consisting of a mix of bridges (some with an iron trough on masonry piers, like Pontcysyllte), tunnels, open-cut channels and cast-iron pipes. About 3,000 people were employed on the project, excluding iron founders and mechanics. Queen Victoria inaugurated the scheme on 14 October 1859 – 'a work which will bear comparison with the most extensive aqueducts in the world, not excluding those of ancient Rome; and one of which any city may well be proud'. Unfortunately Telford did not live to see it.

Back to our tourists: the next day, 22 August, the party set off early from Callander towards Loch Earn. A 14-mile pre-breakfast run took them alongside narrow Loch Lubnaig and, as they approached Lochearnhead, the towering mass of Ben Vorlich – just 22 feet higher than England's highest mountain, Scafell Pike in the Lake District. Southey had seen his first Munro.

> Ere long we reached a very good inn at the broad square head of Loch Earn
> – a house of less pretensions than that at Callander, but in all respects better.

An iron trough on stone piers: the bridge aqueduct in Loch Ard Forest carries water from Loch Katrine towards Glasgow (Geograph/Richard Webb).

Behind it were three hawks, each fastened to its perch, belonging to Sir David Baird, who has probably acquired a taste for hawking in India – certainly the noblest of all field sports, but as certainly the least excusable, because of the cruelties which are daily and habitually practised in training.

During breakfast they noticed a pamphlet for sale: 'Striking and Picturesque Delineations of the Grand, Beautiful, Wonderful and Interesting Scenery around Loch Earn' by one Angus McDiarmid, employee on a local estate. Southey was tempted to purchase a copy from the waiter, but hesitated on discovering that the first sentence 'hath no limitations of sense or syntax'. Telford, wishing to defuse an awkward moment, insisted on adding the choice piece to Southey's collection of curiosities, in spite of the exorbitant price – two shillings and ninepence for just 42 pages, roughly two days' wages for a farm worker, well over £100 in today's money. But according to the waiter Telford's cash was well spent because of the pamphlet's 'high stile'.

Breakfast completed, they continued north through Glen Ogle, 'a fine mountain pass of 8 miles to Killin'. It was the horses' first taste of Highland gradients, part of a military road built in the aftermath of the second Jacobite rebellion of 1745. As they approached Killin the mountain chain guarding Glen Lyon opened up 'like the Mythen-Berg behind Schweitz: four or five of their summits have a strong family likeness'. Southey was becoming entranced, and Telford no doubt gratified by his new friend's enthusiasm.

Loch Earn (Geograph/Sylvia Duckworth).

Loch Tay
(Geograph/Jim Barton).

From Killin they continued 16 miles along the north side of Loch Tay, 'a noble piece of water – like an American river were it not for its apparent want of motion – in its narrowest part I think a mile wide, and perhaps nowhere as much as two'. Southey is enthusiastic about the estate management practised by a local landowner, in spite of the accompanying 'migration':

> The country is very well cultivated. When Lord Breadalbane turned his moun-tains into sheep farms, he removed the Highlanders to this valley. The evil of the migration, if it were so mismanaged to produce any, is at an end, and a wonderful improvement it has been, both for the country and for them. There are marks of well-directed industry everywhere. Flax, potatoes, clover, oats and barley, all carefully cultivated and flourishing; the houses not in villages, but scattered about: and the people much more decent in their appearance, than those whom we saw between Killin and Callander. We met many returning from kirk and carrying their bibles: one old woman, probably unable to go so far from home, was sitting out of doors, reading hers. The children have the vile habit of begging, to the disgrace of the parents, who suffer and most likely encourage it – but they are a healthy and handsome race.

The next night was spent in Kenmore at the far end of Loch Tay – but not very comfortably. The inn was crammed so full that the Rickmans and Miss Emma Piggot were shunted into servants' quarters. Southey and Telford considered themselves lucky to be billeted on a private dwelling nearby where 'the mistress of the house and her daughter, both pleasing and sensible women, received us in the parlour while the room was being made ready'. They escaped the overnight racket that disturbed the rest of the party, but it was the first of many occasions when the two men, newly acquainted, had to share a bedroom.

Next morning, Monday 23 August, they covered the 7 miles to Aberfeldy, 'a place which might properly be called Aberfilthy, for marvellously foul it is. You enter thro' a beggarly street, and arrive at a dirty inn'. Nearby was a bridge:

built by General Wade; but creditable neither to the skill nor taste of the architect. It resembles that at Blenheim, the middle arch being made the principle [*sic*] feature. At a distance it looks well, but makes a wretched appearance upon close inspection. There are four unmeaning obelisks upon the central arch, and the parapet is so high that you cannot see over it. The foundations also are very insecure – for we went into the bed of the river and examined them.

Southey certainly had it in for Aberfeldy, perhaps because of its contrast with the well-managed lands of Lord Breadalbane, perhaps because its bridge offended Telford, the civil engineer. Actually it is nowadays considered one of the best of those built by General Wade, and has been so well maintained over the years that it still carries vehicles, single file, across the Tay.

This is Southey's second reference to Scotland's military roads and bridges, a topic so bound up with Highland history and geography, and of such interest to Telford, that it deserves explanation.

A network of roads, often called General Wade's military roads, was constructed in the Scottish Highlands in the 18th century as part of an attempt by the British government to bring order to rebellious parts of the country – especially those to the west and north of the Great Glen. The first four roads, with a total length of about 250 miles, linked the army garrisons at Fort William, Fort Augustus and Inverness to each other and to the south (see the accompanying diagram), and were constructed between 1728 and 1731 under the direction of General George Wade.

Wade had been sent to Scotland by George I in 1724. He reported that 'more than half of the 22,000 men capable of bearing arms in the Highlands and Islands are ready to create new troubles and rise in arms to favour the Pretender' – a farsighted assessment that presaged Bonnie Prince Charlie's rebellion in 1745. The general was in no doubt that new roads (and about 40 bridges) were needed for rapid and effective deployment of government troops.

 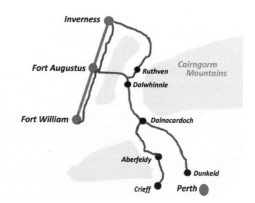

Left: Wade's Bridge over the River Tay at Aberfeldy (Geograph/Richard Murray).

Right: General Wade's first four military roads.

The roads were built by the military to a standard width of 16 feet, reduced to 10 feet where necessary. Construction was limited to the summer months, but even then could be hindered by uncertain weather and the dreaded Highland midges. Each working party consisted of 100 men overseen by two corporals, two sergeants, two subalterns, and a captain. They were generally encouraged by a drummer (but not, apparently, a piper). Camps were established at 10-mile intervals, and the inns which developed near them became known as Kingshouses. The roads, shown on the accompanying map, took the following routes:

Fort William to Inverness (1727)

This road was built along the east side of the Great Glen. The section between Fort Augustus and Inverness proved difficult to traverse in winter, so a major realignment was carried out in 1732, bringing it closer to the shore of Loch Ness. An important feature was High Bridge over the River Spean, about 8 miles from Fort William. The road became familiar to Telford and his colleagues during the construction of the Caledonian Canal nearly a century later.

Dunkeld to Inverness (1730)

Wade started this road at Dunkeld because the existing one from Perth to Dunkeld was reasonably good. The route, which broadly followed that of the modern A9, ran from Dunkeld to Dalnacardoch and Dalwhinnie, an army barracks at Ruthven, and Fort George near Inverness.

Crieff to Dalnacardoch (1730)

Wade judged the existing road from Stirling to Crieff adequate, so ran his road from Crieff to Aberfeldy, where it crossed the River Tay (see earlier photograph). Then on to a junction with his Dunkeld to Inverness road at Dalnacardoch.

Dalwhinnie to Fort Augustus (1731)

This road struck west from Dalwhinnie to Fort Augustus. The most celebrated section, which may still be seen, rises to 2,500 feet over the Corrieyairack Pass. Ironically the road served

The remains of Wade's High Bridge over the River Spean (Geograph/Steven Brown).

*A well-preserved section of General Wade's road over
the Corrieyairack Pass (Wikipedia/Chiswick Chap).*

the Jacobite forces better than the government troops when Bonnie Prince Charlie used it to move rapidly from Fort Augustus towards Edinburgh in 1745.

In 1740 Major William Caulfeild took over the roadbuilding programme from Wade. His name is hardly known to the public; yet he added a further 800 miles to the network, including the following Highland routes:

- Dumbarton to Inverary (1749)
- Tyndrum to Fort William (1752)
- Tarbet to Crianlarich (1754)
- Fort Augustus to Bernera (1755)
- Blairgowrie to Inverness (1756)

Towards the end of the 18th century maintenance of the military roads was increasingly neglected as the Jacobite threat receded, and some sections were abandoned. The rest remained available for commercial and public use, but Wade and Caulfeild had generally gone straight over hills rather than round them, producing steep gradients unloved by carters and civilians in horse-drawn carriages. In 1803 the Parliamentary Commission for Roads, Bridges and Harbours took over responsibility from the military, with Rickman as secretary and Telford as one of the commissioners.

We now return to Southey's account of the tour, which has so far taken us from Edinburgh to Wade's bridge over the Tay at Aberfeldy. The next stop was Dunkeld, 16 miles on:

> As we approached Dunkeld we went thro' the Duke of Athol's [*sic*] plantations, some of the most extensive in the island; indeed I know not if there be any which equal them. He has covered hills which are 1200 feet above the level of the sea, and are so inaccessible in parts, that Mr Telford once asked him if he had scattered the cones there by firing them from cannon. He smiled at the question, being pleased that the difficulty of the enterprise had been thus justly estimated, and he replied that in many places they had been set by boys who were let down from above by ropes.

Southey loved the approach to Dunkeld with its bridge over the Tay, and the cathedral which had 'escaped with less injury than many others from the brutality of the Calvinistic reformers'. And to his delight:

> The bridge is one of Telford's works, and one of the finest in Scotland. The Duke was at the expence, Government aiding him with 5000£. There are five arches, the dimensions of the five middle arches of Westminster Bridge; and besides these there are two upon the land … The Duke wishes to make a better entrance into the town from the bridge; but there is a stubborn blacksmith, whose shed stands just in the way, and who will not sell his pen, thus in a surly doggish spirit of independence impeding by his single opposition a very material improvement.

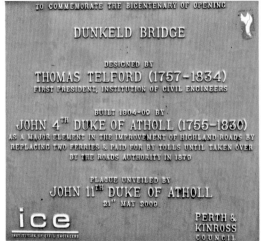

Telford's bridge over the Tay at Dunkeld, completed in 1809 (Geograph/M.J. Richardson).

Commemorative plaque, Dunkeld Bridge bicentenary (Geograph/M.J. Richardson).

Next morning they set out to complete their diversion through the Trossachs. It had cost them three extra days and more than 100 miles of coach travel, but had offered Southey the finest mountain landscapes he would see until they reached the Great Glen. He had plenty to contemplate as they left the 'picturesque country' and headed towards the comparative flatlands of Scotland's east coast.

The alluvial countryside approaching Perth impressed him as highly productive, with more iron ploughs and threshing machines than he had ever seen before; but although he found the Perth 'a good city', the entry into it was 'not favourable' and he summarised the cathedral as a 'poor building'.

Actually, he hardly gave Perth a fair crack of the whip, spending more time on personal health than sightseeing. Calling in on some friends whose house had a fine view of the River Tay and Perth Bridge, he was introduced to a delightful old man by the name of Dr Wood. After 'some cold meat with them, and bottled small beer which was three years old', he asked the medic for professional advice:

> Dr Wood set me at ease concerning one of the tumours on my head which has just begun to suppurate, having been there more than ten years without annoying me before. He says it will discharge itself, and recommends a poultice at night, and some simple ointment on a piece of lint by day. The former part of his advice it is impossible to follow while I am travelling. But I laid in lint and ointment, and must trust to Mr Telford's kindness to apply them: we are generally quartered in a double bedded room.

They left Perth at three in the afternoon, with the aim of reaching Dundee, 22 miles distant, by evening.

To Dundee, Montrose, and Aberdeen

See map on page 69

The coach crossed the Tay on a fine bridge and headed out across the Carse of Gowrie, a 20-mile stretch of fertile lowland along the north shore of the Firth of Tay. Southey describes it as 'a widening vale, evidently alluvial, the land all in open cultivation of the best kind and now in the best season and happiest state'. The fields were full of reapers building small ricks around wooden stakes to provide ventilation and prevent fire; today the Carse is better known as a major area for fruit, including excellent raspberries.

The horses were rested for three quarters of an hour at the village of Inchture, giving the travellers time to inspect a new-fangled weighing machine at a turnpike. But the evening grew dark, raindrops descended, and thunder threatened. Hastening back to the coach they were 'packed in and closed up in less than three minutes', and felt relieved to arrive in Dundee as daylight failed.

This seems a good moment to ask what type of coach they were travelling in. Southey gives no details, simply calling it a 'landau' – a word generally used for a four-wheeled carriage with a team of two or four horses driven by a coachman, or by a postilion mounted on the

Perth Bridge over the River Tay (Geograph/Andrew Abbott).

Looking over the Carse of Gowrie towards the Firth of Tay (Geograph/Paul Taylor).

nearside leading horse. The passenger section has facing seats and a two-part folding top, plus wind-up windows in its doors, to give proper weather protection – especially in a Dundee downpour. The low shell of an open landau makes for maximum visibility by the occupants of their surroundings – and also of the occupants (and their apparel), which is why it is a popular choice for ceremonial occasions.

We know that Southey and his companions totalled a party of seven, including Rickman's wife, two children, and Miss Emma Piggot. The accompanying Daguerrotype image of a landau with its top folded back (and only two of the four horses visible), shows a family group of similar size, and dates from 1849 rather than 1819; even so, it probably gives a fair idea of the setup.

Major new harbour works at Dundee gave Southey his first taste of Telford's work in progress:

A landau ready to depart on a family outing (Wikipedia).

Wednesday, August 25.- Before breakfast I went with Mr T. to the harbour, to look at his works, which are of great magnitude and importance – a huge floating dock, and the finest graving [dry] dock I ever saw. The town expends 70,000£ upon these improvements, which will be completed in another year. What they take from the excavations serves to raise ground which was formerly covered by the tide, but will now be of the greatest value for wharfs, yards, &c. They proposed to build fifteen piers, but T. assured them that three would be sufficient; and in telling me this he said the creation of fifteen new Scotch Peers was too strong a measure.

He is clearly impressed by Telford's wit as well as his works – and adds an unexpected paragraph:

Telford's is a happy life: everywhere making roads, building bridges, forming canals, and creating harbours – works of sure, solid, permanent utility; everywhere employing a great number of persons, selecting the most meritorious, and putting them forward in the world, in his own way. The plan on which he proceeds in roadmaking is this: first to level and drain; then, like the Romans, to lay a solid pavement of large stones, the round or broad end downwards, as close as they can be set; the points are then broken off, and a layer of stones broken to about the size of walnuts, laid over them, so that the whole are bound together; over all a little gravel if it be at hand, but this is not essential.

Southey took away an eclectic mix of impressions from Dundee: a cathedral with pews placed 'as thick as they could be', mystifying the congregation before they were all 'packed up' for the service; a fellow in the streets with a bell in his hand tempting children to gamble for

Left: Victoria Dock, part of Dundee's historic harbour (Geograph/Paul Farmer).

Right: Modern Dundee: the V&A Museum on the waterfront (Geograph/M.J. Richardson).

a lollipop – 'strange that this should be suffered in Scotland'; the hassle of buying a book in one of Dundee's many bookshops, after agreeing to sign his name on a Bank of England £1 note and admitting he was poet laureate. Meanwhile Telford was 'having business with the Provost and other persons touching his great operations which took up the whole morning'. They offered him a formal dinner, but it would have taken the rest of the day. So, time being precious, they set off for Arbroath, watched by a small crowd of onlookers. Southey suspects the bookseller has spread the news that 'my Poetship was to be seen', and was relieved to say goodbye to bonny Dundee without razzmatazz. There were good views of the town as they looked back, and of St Andrews far away across the Firth of Tay.

The 17-mile stage to Arbroath alternately delighted and upset him:

> Large fields and a beautiful harvest, such as to make the heart glad while we looked at it. In many places a poisonous stench from the flax which seems to be extensively cultivated here, linen being the staple manufacture of Dundee. Many neat cottages and houses of a better description, but as usual filthy women with their hair in papers.

The famous Bell Rock Lighthouse, built on a dangerous offshore reef by Robert Stevenson in 1811, was visible from Arbroath – its 'two revolving lights, one very bright, the other less so, being red; they are about three minutes in revolving'. Set against this recent engineering triumph was an ancient abbey, 'which must have been a magnificent building before the beastly multitude destroyed it'. Arbroath centre also had its pluses and minuses:

> The town is very neat and apparently very flourishing: the streets flagged, which they are not at Dundee, to the disgrace of that city, where they have good quarries close at hand. Several booksellers' shops, which indeed seem to be

much more numerous in Scotch than in English towns. And here at Arbroath I saw more prostitutes walking the streets than would I think have been seen in any English town of no greater extent or population.

Rising early next morning they set off for Montrose, a coastal town which Southey preferred to Arbroath thanks to a main street 'broad enough to deserve the name of a *plaza*'. Two miles beyond Montrose they crossed a 'handsome' bridge over the North Esk – but he gives no idea who built it, or when (nor did he mention the engineer responsible for the superb bridge in Perth). I am beginning to wonder why he omitted such important details.

Actually both bridges were designed by John Smeaton, the first Englishman to call himself a professional engineer, and are today celebrated as Grade A structures with commemorative plaques. He completed Perth Bridge in 1771 and North Esk Bridge in 1775, more than 40 years before Southey and Telford crossed them in their landau. Telford was fully aware of Smeaton's prior work in Scotland and must surely have discussed it during long hours on the road and in shared hotel rooms. Was their tour becoming a mutual admiration society, and had Southey decided to mention Smeaton's name as little as possible?

Twelve miles further on they came to Gourdon, a village just south of Inverbervie, where Telford was turning a natural gap between rocks into a proper coastal harbour. The work was almost completed and he had prearranged an important professional engagement:

> We met Mr Mitchell and Mr Gibb, two of Mr Telford's aid-de-camps, [*sic*] who were come thus far to meet him. The former he calls his Tartar, from his cast of countenance which is very much like a Tartar's, and from his Tartar-like mode of life, for in his office of Overseer of the Roads which are under the management of the Commissioners, he travels on horseback not less than 6000 miles a year. Mr Telford found him in the situation of a working mason, who could scarcely write or read; but noticing him for his good conduct, his activity, and his firm steady character he has brought him forward, and Mitchell now holds a post

John Smeaton's fine seven-span bridge over the North Esk (Geograph/Scott Cormie).

of respectability and importance, and performs his business with excellent ability. We left the coach and descended with them to the harbour, the first of the Parliamentary works in this direction. It is a small pier, which at the cost of something less than 2000£ will secure this little, wild, dangerous, but not unimportant port – not unimportant, because coal and lime are landed here from Sunderland, and corn shipped, much being raised in the adjoining country. Mr Gibb has the management of the work. The pier will be finished in about two months, and will shelter four vessels. The basin has been deepened.

This valuable paragraph shows us how Telford was able to carry out his works 'at a distance'. Mitchell, his chief aide, travelled extensively to oversee and report on the many road and bridge projects sanctioned and partly financed by the Parliamentary Commission, of which Telford was a key member (and Rickman the secretary). Gibb (of whom more later) was resident engineer on the project. Such men were crucial guardians of Telford's reputation, and he knew it. At the height of his career, he had a mindbending portfolio of projects under way, not only in Scotland and the rest of Britain, but also in Sweden, where he was heavily involved with the Gotha (Göta) Canal (linking the Baltic to the Skagerrak and the North Sea), a sister to the Caledonian. One of his greatest gifts was the ability to select, promote, and support men like Mitchell and Gibb, regardless of their backgrounds. He could never forget – nor did he wish to – his childhood in Eskdale, and his years as a journeyman stonemason in Edinburgh and London.

Since Telford's involvement with Gourdon Harbour, it has been expanded and renovated several times, most recently in 1960. Today the village's chief attraction to visitors is its strong flavour of a working fishing port, with lobster pots on the harbourside. Close by is a small building housing the excellent Maggie Law Maritime Museum, named after an RNLI lifeboat that saved 36 lives between 1890 and 1930. The local service ceased in 1969 when coverage was transferred to the lifeboats at Montrose and Aberdeen.

Ten miles beyond Inverbervie the touring party reached their next overnight stop, another small town gazing out over the North Sea:

Left: The harbour at Gourdon, a mile south of Inverbervie (Geograph/Stanley Howe).

Right: The Maggie Law Lifeboat Museum at Gourdon (Geograph/Colin Smith).

The harbour front at Stonehaven (Geograph/Andrew Wood).

A long and striking descent upon the smoky town of Stonehaven, with the sea on one side. A fine rocky point to the N. protects the bay from that quarter. The harbour is secured by a small pier, large enough for the place. The inn is on the skirts of the town, as we approached it. Mr Loch, the Marquis of Stafford's agent, left it just after our arrival, travelling south. He is at present exposed to much unpopularity and censure for the system which he is pursuing. Without knowing the merits of the case, his appearance would prepossess me in his favour.

The man referred to, James Loch, was indeed being exposed to unpopularity and censure as the architect of a plan to replace people with sheep in Sutherland – part of the notorious Highland clearances. It seems out of character for Southey to express an opinion based solely on the man's appearance; perhaps Telford enlightened him in the Stonehaven inn that evening.

The next day was given to more immediate concerns:

Friday, August 27: At tea last night, and at breakfast this morning we had Findon [*sic*] haddocks, which Mr Telford would not allow us to taste at Dundee, nor till we reached Stonehaven, lest this boasted dainty of Aberdeen should be disparaged by a bad specimen. The fish is very slightly salted, and as slightly smoked by a peat fire, after which the sooner they are eaten the better. They are said to be in the market (for the most part) twelve hours after they have been caught, and longer than twenty four they ought not to be kept … The haddocks of this coast are smaller than those which are brought to London, or to Dublin, and better; but at the best it is a poor fish, a little less insipid than cod.

15 to Aberdeen. We set out in one of those mists which had a right to wet R. [Rickman] and myself, as Englishmen, to the skin; so we were all packed in the inside, and the landau was closed.

Strangely, this is the first mention of Rickman since they left Edinburgh; and his wife, children, and Miss Emma Piggot might as well have stayed in the capital for all the attention Southey has given them. Yet they were travelling together, packed like Finnan haddock in a fish basket and overnighting in a wide variety of accommodation, some of it extremely challenging.

On reaching Aberdeen, Rickman and Telford set off with the estimable Mitchell to inspect a road and some bridges about 30 miles to the west, leaving Southey and the rest of the party in the care of Gibb (superintendent of Aberdeen's new harbour works) and Haddon, the city's provost. After dinner Gibb escorted them on a tour that included the university's Marischal and Kings colleges. Old Aberdeen, with a population of about 1,500, had 'something of a collegiate character – an air of quietness and permanence'. But all was 'life, bustle, business

Union Street, Aberdeen (Geograph/Colin Smith).

and improvement' in the New City, almost on a scale with Edinburgh: 'Union Street, where our hotel stands, is new, and many houses are still building – the appearance is very good, because they have the finest granite close at hand': Aberdeen's granite, and the stonemasons who fashioned it, were renowned throughout Europe.

Ever since leaving Perth Southey had been relying on Telford to 'dress his volcano', but he now needed a doctor's advice again. Perhaps unfortunately, Telford had told him a story about a man who broke a leg in the Highlands and was carried to the nearest town by two locals, who started arguing about claiming a reward. The man's servant, fluent in Gaelic, heard one insist that 'we must ask enough, for it is not every day that an Englishman comes here and breaks a leg'. We may imagine Southey's anxiety as he approached an Aberdonian doctor of unknown reputation and competence, who

> prest much of the subaceous substance out of the tumour, talked of enlarging the orifice, of the danger of inflammation communicating to the dura mater and so on; of lying by for three days and he affirmed that the contents could not be carried off by suppuration where there was no vitality. I have very often wished myself wholly ignorant of anatomy and nosology, having repeatedly felt that a smattering in these things serves only to disquiet us, both for ourselves and others, with apprehensions generally ill-founded, but specious. In this case however I knew enough to perceive that this man was presuming upon my ignorance, and endeavouring to deceive me by a mixture of nonsense and falsehood.

He beat a hasty retreat, begging Gibb to transfer him to the best surgeon in town. The resultant Mr Kerr, whom they found 'reading Aristotle in Greek', was a very different type of medic: previously attached to the army and a disciple of non-intervention, he told the patient that Telford's continuing kindness was all he needed, and charged a modest fee. Southey left feeling 'relieved from no inconsiderable anxiety'.

When Rickman and Telford returned from their inland expedition, they all went to the harbour. Southey finally brings himself to mention John Smeaton, but in a slightly backhanded way:

> The quay is very fine, and Telford has carried out the pier nine hundred feet beyond the point where Smeaton's work terminated. This great work, which has cost 100,000£, protects the entrance from the whole force of the North Sea. Gibb has the superintendance of the work. A ship was entering under full sail – The Prince of Waterloo – she had been to America, had discharged her cargo at London, and we now saw her reach her own port in safety – a joyous and delightful sight. The Whalers are come in, and there is a strong odour of train oil, which would rejoice the heart of a Greenlander, and really even to us it was perfume after the flax … Coal and lime are brought to this country from Sunderland, and the lime is carried many miles inward for dressing the land.

Gibb was actually a talented civil engineer in his own right – as he must have been, to superintend such a huge project. He had worked on docks in Edinburgh and Greenock, and in 1811 Telford offered him the job of resident engineer in Aberdeen at a salary of £250 a year. He then spent six years extending and fortifying the harbour, repairing the south pier, and constructing a breakwater, north pier, and new dock walls. He was the first engineer to use steam dredgers in Scotland (anticipating Telford's need for them on the Caledonian Canal). By the time Southey met him in 1819 he had contributed to numerous projects, including the harbours at Gourdon, Peterhead, Banff, Cullen, and Nairn.

Now and then Southey lightens his account with amusing observations of Scottish traditions and (as seemed to him) eccentricities. In Aberdeen he notices the 'town cryer' summoning people not by bell, but with a drum; a beggar in the street reading his bible;

Aberdeen Harbour (Geograph/Nigel Corby).

Peterhead Harbour (Geograph/Mike Pennington).

*Bridge of Don, Aberdeen,
by Thomas Telford
(Geograph/G.Laird).*

bread and butter made with only the soft part of the bread, 'a bad practice'; Finnan haddock monotonously eaten at breakfast and at tea; whisky decanters of manifold shapes and sizes including choppins, mutchkins, and gills; and sash windows lifted by two brass handles. To offset any perceived Sassenach bias he is happy to add a compliment:

> The Scotch regard architectural beauty in their private houses, as well as in their public edifices much more than we do; partly perhaps because their materials are so much better. For as for making fine buildings with brick, you might as easily make a silk purse out of a sow's ear.

The next morning, 29 August, Telford set off early with Gibb and Mitchell to inspect harbour works in progress at Peterhead, with an arrangement to rejoin the others in Banff the following day. Southey walked with Rickman to Aberdeen Old Town and the historic bridge over the River Don – 'well placed, having rocky abutments on both sides'. They could hardly have foreseen that an Act of Parliament six years later would allow the city council to replace it with a new bridge, or that Telford, with Gibb's help, would design and build it. Today the five-span granite 'Bridge of Don' is listed as a historic structure, and gives its name to a suburb on the north bank of the river.

At four in the afternoon Southey, Rickman, and company left Aberdeen, said goodbye to the east coast road they had followed all the way from Perth, and headed inland.

To Banff, Elgin, and Grantown-on-Spey

See map on page 69

Possibly to their surprise (and certainly mine) for the first 6 miles out of Aberdeen towards Banff the road runs close to a canal. Southey has mixed feelings:

> The canal is a losing concern to the subscribers; and Mr Haddon complains that it draws off water from the Don, to the hurt of his mills; he is a great

manufacturer, employing 3000 persons! It is however a great benefit to the country, and no small ornament to it, with its clear water, its banks which are now clothed with weedery, and its numerous locks and bridges, all picturesque objects and pleasing, where you find little else to look at.

The Inverurie Canal, now generally referred to as the Aberdeenshire, has been so overgrown by history that few people have heard of it, even in Scotland. It may have pleased Southey's eye, but not the pockets of its shareholders – a problem that dogged it from 1805 to its closure 49 years later. Designed by the famous John Rennie, it ran 18 miles from Aberdeen to Port Elphinstone, just short of Inverurie, with 18 locks big enough for boats 57 feet long. A wide variety of cargo was carried and passenger services ran twice a day in summer and once in winter – assuming the canal was ice-free. A sea lock into Aberdeen Harbour was built in 1834, easing transshipment of goods to seagoing vessels. But like canals throughout Britain, the Aberdeenshire fell victim to the railway revolution: by 1845 the canal company was negotiating a sale to the Great North of Scotland Railway, which closed the waterway nine years later and laid tracks over most of its course. Today a few remnants are all that survive from its not inconsiderable heyday; some may still be seen from the river and the canal walk that starts in Port Elphinstone.

Southey's next place of interest was their stop for the night, Old Meldrum:

> a small village, or townlet, consisting chiefly of a plaza in which there is a town-house. The Inn is comfortable – more so than if there were more pretensions about it. We had a fire for the first time on our journey – the material was peat. It was well that beds had been bespoken; two persons would have passed the night here, unless we had thus pre-engaged the quarters; and then it would have

This remnant of the Aberdeenshire Canal was subsequently used as a mill leat (Geograph/Bill Harrison).

been difficult to have housed us. The tea kettle had a stand, as if for a lamp, but it contained a heater.

Unfortunately his bedroom turned out to be as small as a ship's cabin. A small jug was placed under the window to hold it up, and a clock that chimed 'with eight sweet bells every quarter' destroyed his sleep. On the positive side, the breakfast was as generous as usual in Scotland, with Finnan haddock (yet again), eggs, sweetmeats, honey (and presumably bread), which set them up for the day's journey.

The 16 miles to Turreff [Turriff] began over waste land but soon showed signs of improvement. Southey finds the people of Aberdeenshire adept at enclosing ground with stone walls but, where stone is absent, negligent about using any other material. As a result cattle 'must always be watched', and there is no shelter for grain, 'greatly as it must be wanted'. Turreff was larger but more straggling than Old Meldrum, with one church in ruins and another stranded in a kitchen garden. Its best house belonged to a lawyer – 'proof how the profession flourishes in Scotland' – and the small town boasted no fewer than three watchmakers.

Their landau soon covered the 10 miles to Banff, where the travellers had their first view of the Moray Firth, 'open sea, not as we had hitherto seen it, grey under a sunless sky, but bright and blue in the sunshine'. They crossed the Deveron river by a 'good bridge of seven arches by Smeaton', and Southey espied a church with a spire acting as a landmark for shipping, and the smaller town of Macduff about a mile across the bay. The day's major event was about to start:

> Here we rejoined Mr Telford, Mitchell and Gibb, and went with them to the pier, which is about half-finished, and on which 15,000£ will be expended to the great benefit of this clean, cheerful, active town … The pier was a busy scene – handcarts going to and fro on the railroads, cranes at work charging and discharging huge stones, plenty of workmen, and fine masses of red granite from the Peterhead quarries. The quay was almost covered with barrels of herrings, and women employed in salting and packing them. They were using much more salt than was necessary, and it was mortifying to learn that this is done because they are allowed an exemption from the duty on salt. The barrels are sold by weight, and the more salt the worse for the fish, but the better for the seller. It seems that this exemption which is so well intended, and at first appears so just and unexceptionable, gives occasion to great frauds, smuggling, and evil in many ways.

About 30 fishing vessels were sighted early next morning as the landau struggled up the hill out of Banff, travelling west. They passed through Portsoy, 'a neat, thriving little place, where a good proportion of the houses have gardens, and several are prettily clothed with creepers, or fruit trees'; and entered the territory of Colonel Grant, 'the head in possession of that family, Lord Seafield being insane'. Southey seems a little obsessed by the Grants, and especially the claimed antiquity of their clan, perhaps because they were invited to the Colonel's mansion for a magnificent breakfast – 'fresh herrings of the finest kind being added

*Left: Banff Bridge, built by Smeaton in 1779 and widened in 1881, carries today's
heavy traffic over the River Deveron (Geograph/Anne Burgess).*

*Right: Banff Harbour was extended by Telford in 1819 using red
granite from Peterhead quarries (Geograph/Stanley Howe).*

here to the usual abundance of good things'. They then walked about a mile to the port of
Cullen:

> The works here, of which the whole expence will be about 4000£, are in such
> forwardness that the pier at this time affords shelter: and when I stood upon it at
> low water, seeing the tremendous rocks with which the whole shore is bristled,
> and the open sea to which it is exposed, it was with a proud feeling that I saw the
> first talents in the world employed by the British Government in works of such
> unostentatious, but great, immediate, palpable and permanent utility. Already the
> excellent effects are felt. The fishing vessels were just come in, having caught about
> 300 barrels of herrings during the night. All hands were busy. Some in clearing
> from the nets the fish which were caught by the gills; some in shovelling them
> with a long and broad wooden shovel into baskets; women walked more than
> knee deep into the water to take these baskets on their backs, while under sheds
> erected for a protection in hot weather, girls and women out of number were em-
> ployed in ripping out the gills and entrails, some others in strewing salt over them,
> and others again in taking them from the troughs into which they were thrown
> after this operation, and packing them in barrels. Others were spreading the nets
> to dry; the nets are of a dark brown colour, dyed thus by a decoction of alder bark
> in which the thread is dipped to preserve it. Air and ocean also were alive with
> flocks of sea fowl, dipping every minute for their share in the herring fishery.

The herrings were chiefly sent to the West Indies and the Mediterranean. Other species
were regarded as enemies: seals, because they indulged in salmon – the sealskins were
stripped off whole and inflated for use as buoys for the nets – and dogfish, up to 2 feet long,
because they tore the nets.

Southey is fascinated by the rapid development of the herring fishery:

> A fishing village had already grown up on the shore, for the white fishery, and thus there was a race of fishermen in existence upon the spot when the herrings were discovered. If the present spirit continues, the Dutch will soon be rivalled, and probably exceeded in this branch of industry. There are sometimes 300 vessels at once employed in the Moray firth … The whole line of coast is in a state of rapid development, private enterprize and public spirit keeping pace with national encouragement, and it with them. Government is to blame for not making its good works better known … The money which it bestows upon harbours arises from the remainder of the rents of the forfeited estates … However much the money may have been misapplied during a long series of years, by those to whom it was entrusted, the remainder could not have been better applied. Whenever a pier is wanted, if the people or the proprietor of the place will raise half the sum required, Government gives the other from this fund, as far as it will reach. Upon these terms 20,000£ are expending at Peterhead, and 14,000 at Fraserburg [Fraserburgh], and the works which we visited at Bervie [Inverbervie] and Banff, and many other such along this whole coast would not have been undertaken without this help from the Government; public liberality thus directed inducing individuals to tax themselves liberally, and expend with a good will much larger sums than could have been drawn by them from taxation. At Cullen we took leave of that obliging, good-natured, useful and skilful man, Mr Gibb.

The Scottish herring industry, which Southey was witnessing in its early stages, grew with lightning speed in the 19th century, thanks in no small measure to the development of new harbours. Prior to 1800 the boats were small, wooden, and open, with one or two masts, and could be dragged up the beaches. Inexpensive to build, they were repaired by the fishermen themselves. Unfortunately, however, a self-employed Scottish crofter/fisherman lacked the

Cullen Harbour (Geograph/ Alan Hodgson).

resources to participate in the highly organised European industry, which was dominated by the Dutch with huge boats that went to sea for months at a time and preserved their catches on board using a secret 'Dutch Cure'.

However by the end of the 18th century the long period of neglect and mismanagement of British fisheries was coming to an end. The British Fishery Society, founded in 1786, had helped lay the foundation of a large and efficient herring industry by developing fishing ports on the west coast of Scotland, beginning at Tobermory in 1787 and Ullapool in 1788. Telford started a lifelong connection with the society in the 1790s, giving free advice on many projects including a new settlement and harbour at Wick in the far northeast, completed in 1811. This gave a tremendous impetus to the east coast herring fishery, which was further advanced by the discovery of a rival 'Scotch Cure' in 1819.

The British government started subsidising large boats, with an additional bounty on herring sold abroad. The fish were caught in huge quantities off Scotland's east coast in winter and spring, and off the north coast and Shetland in summer. Boat crews followed the fish up the coast, backed up on land by itinerant teams of 'herring girls' who gutted, salted, and packed the 'silver darlings' in barrels. By mid-century tens of thousands of boats were engaged in the trade, spurred on by new railways that gave fast access to markets at home and abroad. At the peak of the herring boom in 1907, two and a half million barrels of fish were cured and exported, mainly to markets in Germany, Eastern Europe and Russia. So by the 1920s the fishery, unregulated and heavily overexploited, was unsurprisingly drifting towards collapse.

After they left Cullen the touring party headed 12 miles inland to Fochabers, where Mitchell's assistant arrived in his personal gig (a light two-wheeled carriage drawn by one horse), and Southey received a welcome letter from home. They walked into the Duke of Gordon's park to see a small hump-backed zebu (an Indian cow), and gazed at his 'great ugly house'. Two fox-brushes graced (or perhaps defaced) the dining room of the local inn, accompanied by a print of the 'late notorious Duchess of Gordon in her youth, and another taken toward the close of her life'.

Southey seems to have it in for the Gordons, and especially His Grace:

> The Duke lets the salmon fishery for 7000£ a year (a sum almost incredible) with a stipulation that he himself is to be supplied at sixpence per pound. But he has made no such stipulation for the people of the place; they can only obtain it as a favour and at the price of a shilling. This is a great injustice and vexation, growing out of a feudal right, in the origin of which no such wrong could possibly have been contemplated.

Once again the English poet alternates between admiration and disdain for wealthy and powerful Scots, in this case Colonel Grant in Cullen and the Duke of Gordon in Fochabers. Could it have anything to do with being entertained at a magnificent breakfast by one, and denied access to a 'great ugly house' by the other?

Fochabers itself gets only the most cursory of mentions: it has 'a bridge over the Spey, of five arches, the middle one 90 feet high'. Yet this was a Telford design – just the sort of structure Southey would be expected to admire, and describe in some detail. He was probably having a bad day.

Telford had completed the handsome sandstone bridge in 1806 – 13 years before Southey saw it. Like the ancient packhorse bridge across a tributary of the Spey at Carrbridge (mentioned earlier), it would be partially swept away in the 'Muckle Spate' of 1829. Today's Old Spey Bridge is an unhappy combination of Telford's surviving stone arches and a cast-iron replacement concocted by an unrecorded engineer in 1854.

Their next stage covered the 12 miles to Elgin, a city that Southey, in his present mood, finds unattractive: the sight of a ruined cathedral makes him 'groan over the brutal spirit of mob-reformation'; a climb onto the surviving roof of the Chapter House causes giddiness, a symptom of his 'declining life'; the city has an appearance of decay about it; and he is irritated by a bell rung at eight o'clock and an 'abominable drum beaten at nine'.

Bridges and roads in Strathspey were on the extensive menu next day, and they left the hotel at six in the morning. The plan was for the ladies and children to take the landau to Nairn while the men went on an engineering detour, Telford and Mitchell in the latter's gig, Southey and Rickman in a hired one. The men breakfasted in Rothes, at a comfortable little inn

> without any pretensions to finery; and there is in the garden, to borrow a clean word from France (a country which has little cleanliness to lend) that *Commodity* the want of which was formerly the reproach of Scotland, but is no longer so, for, except at Aberfeldy, I have found it at every place where we have stopt.

The mind boggles.

Two miles beyond Rothes they came to a high point of the tour: Telford's wonderful Craigellachie Bridge over the River Spey. The approach road:

Left: Swirling waters: the Old Spey Bridge at Fochabers (Geograph/Anne Burgess).

*Right: Telford in a gig with Rickman's son just visible
(sketched and tinted from a 19th-century image, artist unknown).*

brings the traveller by a short tour to the two short turrets at the entrance of the bridge: on this side they are merely ornamental, on the other their weight is necessary for the abutment. The bridge is of iron, beautifully light, in a situation where the utility of lightness is instantly perceived. The span is 150 feet, the rise 20 from the abutments, which are themselves 12 above the usual level of the stream. The only defect, and a sad one it is, is that the railing for the sake of paltry economy is of the meanest possible form, and therefore altogether out of character with the rest of the ironwork, that being beautiful from its complexity and lightness. But this farthing-wisdom must now appear in everything that Government undertakes; and thus the appearance of this fine bridge has been sacrificed for the sake of a saving, quite pityful in such a work. Mr T. undertook to finish the bridge in twelvemonths: it was begun in June, and opened in the October following. The ironwork was cast at Plas Kynaston, and brought by the canal over his great aqueduct at Pontey-Syllty [Pontcysyllte]. The whole cost of the bridge and approaches was £8,200.

Craigellachie displays Telford's unequalled mastery of cast iron. He dazzles us – as he did his contemporaries – with iron tracery so light and elegant that it appears suspended over the river like a delicate butterfly wing. Fortunately, as Southey notes, his trusted colleague William Hazledine of Plas Kynaston, one of the finest ironmasters in the world, was able to repeat the collaboration first established on the Pontcysyllte Aqueduct in northeast Wales. And when the cast-iron sections had arrived on site via canal, sea and road, Hazledine's foreman William Stuttle and two assistants were despatched to Speyside to superintend construction. Craigellachie was the only bridge across the Spey to survive the catastrophic Muckle Spate of 1829; Telford had understood that the river's notorious liability to heavy flooding must

Craigellachie Bridge by Thomas Telford, completed in 1815 (Geograph/Euan Nelson).

The turrets of Craigellachie (Geograph/Euan Nelson).

be countered by raising the structure far above the river bed and crossing it with a single span. An iron plaque with the legend 'Cast at Plas Kynaston Ruabon Denbighshire 1814' still celebrates Craigellachie, the oldest surviving iron bridge in Scotland.

Telford, Mitchell, Rickman, and Southey were now heading up Strathspey towards Grantown, but took a short detour at Bridge of Avon to accept hospitality at Ballindalloch Castle from Mr Macpherson-Grant, MP for Sutherlandshire. The castle, 'pearl of the north', had been the family home of the Macpherson-Grants since 1546, and both the Spey and its

Ballindalloch Castle (Geograph/Andrew Tryon).

Avon tributary flowed through its grounds. Telford had previously saved the family's Spey bridge from ruin, no doubt making him a welcome visitor. Southey describes their host as 'a sensible and agreeable man of six feet two', who showed them a portrait of General Wade – presumably a hero in present company – wearing a blue velvet robe over his breastplate, and a wig: 'the countenance mild and pleasing, by no means deficient in intellect, but not indicating a strong mind'. It is hard to believe that a weak mind could have taken responsibility for 250 miles of military roads across the Highlands.

They walked around the grounds and then went in for a challenging lunch:

> For the first time I tasted sheep's head broth, which is rather better than hodge-podge; the flavour of the burnt wool hardly differs from what might be given by burnt bread, and might be better obtained by burnt cheese. The head itself was a separate and ugly dish, which I did not taste, supposing it to be as bad as calve's head. We closed all with a bottle of claret, and took our leave at four, well pleased with our host, his family and his habitation. He is a very domestic and useful man; and has been the great promoter of the roads in this part of the country.

It seems that the remaining 12 miles to Grantown-on-Spey were at least partly on a road constructed by Telford, and Southey summarises the engineer's approach to roadbuilding – 'After the foundation has been laid, the workmen are charged to throw out every stone which is bigger than a hen's egg … A clear set of principles regarding drainage ensures that the work is permanent in all its parts.' The four men in two gigs now experienced at first hand the quality of a Telford surface, and since Southey makes no complaint they presumably found it comfortable.

Not so the weather: it rained heavily and blew hard for most of the stage, and whereas the gigs' aprons protected their feet and knees, and the umbrellas their heads and shoulders, they were 'wetted in tail'. They must have envied the ladies and children their landau ride towards Nairn. The storm lasted an hour and a half, but abated as they approached General Wade's bridge on the outskirts of Grantown. Southey, no longer surrounded by Wade enthusiasts, enjoys another tirade against the military engineer's approach to bridgebuilding:

> Like all the General's bridges, it was miserably constructed, and had a tremendous rise: this evil has been corrected, and the bridge itself preserved from the ruin which must otherwise speedily have befallen it – but there are tremendous cracks in two of the arches.

Could it be that the iron straps seen in the accompanying photo have been holding the structure together for well over two centuries?

The night was spent in Grantown-on-Spey which (to my surprise) Southey describes as 'dull and uniform'. Equally surprising is that their inn, the Grants Arms, turned out to be run

*Left: Old Spey Bridge near Grantown-on-Spey,
built by General Wade in 1754 (Geograph/John Lord).*

Right: Grant Arms Hotel, Grantown-on-Spey (Geograph/Paul Anderson).

by an ex-army man and his wife – 'a forward, vulgar, handsomish woman from Portsmouth' – who seemed to hold Grantown in contempt, were out of food when they arrived, had decorated the place with vulgar prints, and had wasted money on good furniture quite inappropriate for an establishment 'no better than a village alehouse'. Clearly their 'small inn' cannot have been the famous Grant Arms in Grantown, which dates from 1765 and enjoyed a visit by Queen Victoria in 1860. But surely the poet laureate and his companions would have preferred rooms in a spacious and comfortable hotel? Perhaps there had been a muddle between 'Grants' and 'Grant'; in any case they found themselves stuck in a decidedly inferior establishment, saved only by a fine peat fire that dried their wetted tails, and half a mutchkin of whisky that wetted their whistles before bedtime.

To Forres, Nairn, and Inverness

See map on page 69

The next morning, 2 September, they left Grantown-on-Spey at seven and travelled north to meet the ladies and children in Nairn. The first 14 miles to the Fairn-ness [Ferness] Inn was by a military road driven through wasteland – heath growing in peat, gravel below, small lochs, and steep Wade bridges over burns. The 'excellent road', now maintained by the Parliamentary Commission, passed some miserable human habitations

> as black as peat stacks in their appearance, the peat stack in reality generally forming part of the edifice – peat stack, peat sty, and peat house being all together: the roofs are also covered with turf, or peat, on which grass and heather grow comfortably, and probably the better because of the smoke which warms the soil. But great quantities of wood are brought from Strathspey by this line to the coast. Mitchell has sometimes met thirty carts together.

Breakfast was taken at the Ferness Inn, established on an otherwise bleak military road to please the roadmakers. Our travellers had

> a good fire and a comfortable meal. Salted herrings were part of our fare … The woman told us she had just heard that the fishing vessels from Burgh-head had lost their nets, so many herrings having been caught in them that they were not able to lift them out of the water. Upon enquiry at Nairn we found that this had really happened. As soon as the herrings die they sink, and then the net is lost. Many vessels have thus lost their nets lately, some have drifted to Fort Rose and there been recovered; and the sea is said to be infected with the stench of the dead fish.

So the breakfast talk was of stinking herrings from the Moray Firth, not sparkling salmon from the River Findhorn running nearby.

Ferness Bridge over the Findhorn, completed by Telford in 1817, is a Category A listed structure described by Historic Scotland as 'a 3-span granite rubble bridge with 3 shallow segmental arch rings, the centre slightly larger than the outer arches, all dressed granite. Dressed and rusticated triangular cutwaters; string course and parapet; slightly splayed approaches. 150 feet span'. It is also single track, as today's motorists can confirm. Curiously, Southey describes it simply as 'a good plain bridge, built by the Commissioners'. He had probably been so beguiled by the magic of cast-iron Craigellachie, visited the day before, that he found a three-span granite bridge, even an excellent one designed by Telford, underwhelming.

Their next stop was Forres, 12 miles on through relatively 'improved' country. Approaching the town they encountered two splendid carriages, each with four horses, and realised that they belonged to Leopold, a German prince who had married an English princess and would later become the first king of the Belgians. Sadly his wife had recently died in childbirth. The bereaved prince was a popular figure in Britain, he was on tour, and Forres was in commotion:

Temporarily closed for repairs: Telford's single-track Ferness Bridge over the River Findhorn (Geograph/John Lord).

Nelson's Tower, Forres (Geograph/nairnbairn).

Two bells, the whole peal of one kirk, were working away, ding-dong, ding-dong, and one in another went ding-ding – a comical presbyterian attempt at bell ringing – and from a sort of summer house on a wooded hill, called Nelson's Tower, one gun was fired as fast as it could be loaded. In the midst of this bustle we stopt at Loudon's Hotel. The Prince was taken to another, and much worse house, by the Marquis of Huntley, for this notorious reason, that it is kept by a woman who – is kept by him.

Standing high on Cluny Hill, Nelson's Tower looks down with considerable authority over Forres. It was funded by public subscription and built as a memorial to Admiral Lord Nelson. The foundation stone was laid in 1806, and the Forres Volunteers fired three volleys into the air before marching down the hill for a celebration dinner in the town. The building was completed in 1812 and a Trafalgar Club formed for annual celebrations of the famous naval victory. It hardly seems fair of Southey to describe all this in terms of a summer house – why not a lighthouse?

Southey was received and flattered by Lady Cumming Gordon at Eleanor Castle in Forres, where 'Heaven knows how many persons came in by one, by two, and by three, to look at me and repeat their civilities'. Afterwards they drove 11 miles to Nairn, crossing the heath 'where the Witches met Macbeth', and rejoined the ladies and children. He found a small parcel waiting for him in the hotel, containing a diploma from the Literary Society of Banff and a letter from the secretary welcoming him as an honorary member. It was further evidence that his reputation as a poet laureate on tour was beginning to run ahead of him. But his experience in the hotel that evening was far less attractive:

> When Mr Telford and I retired to our double-bedded room, we heard a great knocking over head, and ringing the bell to request that the persons above might be desired to be less noisy, were told that it was only the Masons, who had just done, and were going away. We went to bed, but the knocking and other unaccountable noises continued at intervals until one o'clock; T. then exclaimed that this was too bad; and as it was impossible to sleep we began to talk about it. I who, when the Chamber-maid spoke of the Masons, understood that word in its usual meaning … supposed they were employed up on the roof of the house … But T. laughing at my mistake, told me the Freemasons were holding a lodge upstairs. So it proved to be … immediately over our heads, at midnight, the aspitants were going thro' what in their Diplomas are called "the great and tremendous trials" of initiation. Whether I actually heard a great cry or only dreamt it I am not certain, but I think I heard it.

Were the travellers responsible for their own hotel bills, or had they been settled in advance? Southey fails to tell us, or whether they were offered a rebate the next morning by a suitably contrite manager.

Southey says little else about Nairn. This is surprising given his attention to other coastal towns, because Nairn was in sore need of a proper harbour for fishing boats. It was located

*Nairn's harbour piers, recommended by Telford, extend
out into the Moray Firth (Geograph/Colin Smith).*

at the mouth of a sizeable river whose sandy estuary was shallow and unstable. It seems that Telford was first consulted about it in 1818, and submitted a plan for a harbour and wharf, with piers extending straight out into the Moray Firth. His recommendations were accepted, and although the works were subsequently damaged by floods the present harbour owes its basic shape to him.

Next morning the party covered the 7 miles from Nairn to Fort George. It was, and still is, an extraordinary fortress. Ordered by George II to pacify the Scottish Highlands in the aftermath of the Jacobite rising of 1745, it replaced an earlier one in Inverness that had been built after the 1715 rising only to be blown up by the Highlanders. With sheltered waters below its walls, the second Fort George could be supplied by sea in the event of a siege. In fact it has never been attacked, and is still used by the army. The fortifications, based on a star design, remain virtually unaltered and are nowadays open to visitors. Originally the depot of the Seaforth Highlanders and later of the Queen's Own Highlanders (Seaforth and Camerons), it has recently been home to the Royal Irish Regiment and, as of 2007, the Black Watch.

What did the poet laureate make of this extraordinary place, completed 70 years before his visit?

The Fort is a specimen of regular fortifications which may be regarded without one melancholy reflection, no gun having ever been fired in anger either from or against it. Not that I mean to speak of it with contempt – far otherwise. It was necessary when it was built, and is useful now. We ought to have some such place where Officers may make themselves practically acquainted with fortifications. It will hold 1900 men, but at present has scarcely more than a 19th part of that number. The interior a good deal resembles one of the Inns of Court. In some of the covered ways, the weeds which grow between the palisade and the wall, have reached more than double the height which they would have attained in another situation; shooting up toward the air and light, they are full ten feet high.

Noncommittal and rather poetic – no shots fired, magnificent weeds. And his next sentence? '13 to Inverness.'

Arriving in Inverness, Southey returns to less sensitive topics:

We met a large party of men and women, chiefly sailors, with a funeral. The body was borne on a hand bier, covered with an old velvet pall. Cawdor Castle

Above: Fort George from the air (Geograph/Graeme Smith).

Left: Fort George on the ground: soldiers' quarters (Geograph/Michael Garlick).

was in sight; and we passed by Stuart Castle, a ruin of the Flemish kind. The road is a military one repaired by the Commissioners. It was Fair Day when we arrived, and the streets were filled, mostly with women; they appear never to wear hats or bonnets – but either a white cap, or a white handkerchief over the head; or the head is bare.

His fascination with women's heads may have been aroused by his own, which was again giving trouble:

Here I sent for a surgeon, Kennedy by name, who was an acquaintance of Mr Telfords. His report was not a pleasant one; proud-flesh has formed within the cavity, and he means to apply a mercurial ointment, but he says there is no occasion for confinement or rest. He is said to be a good surgeon, and has had long experience in the army – yet he talked of correcting my blood by Epsom Salts! As if a tumour of ten years' standing had anything to do with the present state of my blood, or as if Epsom Salts could alter that state. If I were to remain here he would apply caustics; this ointment is escharotick, to produce the same effect more slowly. I took only half his dose, because no medical man will be persuaded that half the quantity of any medicine which would be required for other men, suffices for me.

Poor Southey: whenever he arrived in a major town or city he sought a respected doctor, only to receive advice he could not trust. His 'cavity', which he also called his 'volcano', was causing him continual anxiety. It had erupted soon after leaving Edinburgh, and he was now halfway through a demanding six-week tour with friends he admired, dependent on the daily ministrations of an engineer who was kind enough to dress the wound.

They left Inverness next morning and crossed the Caledonian Canal just below the four Muirtown locks. There was no time on this occasion to inspect the works; instead they called at the house of Mrs Davidson, whose husband Matthew, Telford's principal lieutenant at the northern end of the canal, had very recently died – a bitter blow to everyone concerned. Southey reminisces about the man's extraordinary character, and notes that Mrs Davidson placed a basket of excellent gooseberries in their landau, together with 'some of the finest apricots I ever saw or tasted, which have grown out of doors ... her husband was fond of cultivating his garden'. So ended the most poignant 24 hours of their tour.

To Dingwall, Bonar Bridge, and Dornoch

See map on page 69

Highland geography between Inverness and Dornoch is confusing to almost everyone ex-cept a local. The first time we encountered the Black Isle and the sequence of three firths – Beauly, Cromarty, Dornoch – I well remember a sense of disorientation as we wrestled with scenic transitions from land to water and back again. I hasten to add that it was all very beautiful.

Southey's journal, too, is confusing at this stage of the tour. His account contains no maps or illustrations (apart from portraits of himself and Telford); anyway, most modern travellers prefer the greatly improved A9 to the limited road network of 1819. In a nutshell, the party had to skirt around the firths rather than cross them by modern bridges; and they sometimes diverted to investigate something special. I have had to read some of his pages several times.

The next part of their journey took them to Lovat Bridge at the head of the Beauly Firth, where they diverted 4 miles to the southwest along Strathglass to see several beauty spots and a new road (the present A831), 'one of the most remarkable of them, for the difficulty of constructing it … in many places it is cut in the cliff, and in many supported by a high wall, a work of great difficulty, labour, and expense'. And then back again:

> Lovat Bridge, to which we returned, cost 8800£. The foundations were expensive, and the stone was not at hand. It is a plain handsome structure of five arches, two of 40 feet span, two of 50, and the centre of 60. The curve is as little as possible. I learnt in Spain to admire straight bridges; but T. thinks there always ought to be some curve, that the rain water may run off, and because he would have the outline look like the segment of a larger circle, resting on the abutments ...

Southey omits to say that the bridge was actually one of Telford's, completed in 1814 as a major element in his northern Scottish road improvement programme. A few miles further on they would come to his Conon Bridge, a similar design but about £3,000 cheaper thanks to easier foundations and abundant local materials.

They stopped at Beauly for 'a decent dinner of salted ling, eggs, mutton chops, and excellent potatoes, with ginger beer, and good port wine at what appeared no better than an English alehouse'. Next came Dingwall, a small town for which Southey shows little regard. He notices a 'ridiculous obelisk' opposite the kirk, held up with numerous iron cramps and bars; and learns that Davy, a strange local employed by Telford, is so fond of whisky that, for the sake of his family, he cannot be trusted with the money he earns paving the streets of Dingwall – 'a mark of civilisation that ill accords with the general aspect of the place'. Next morning they set off for Invergordon, 15 miles distant, skirting Cromarty Firth and passing an estate

> on which Sir Hector Monro expended the whole wealth which he acquired in India, so that he was obliged to go to India again and make a second fortune for the purpose of enabling him to live upon it. The spoils of the East have seldom been better employed, than in bringing this tract which was then waste ground, into a good state of cultivation. There are extensive plantations on the hills behind the house, and some odd edifices on the summits which he is said to have designed as imitations of the hill-forts in India. One of them appeared like a huge sort of Stonehenge; but we saw it only from a distance.

Cromarty Firth (Geograph/ djmacpherson).

Invergordon, like Dingwall, had few charms for Robert Southey, but was important for the ferry that connected it to the Black Isle, whence another ferry sailed to Inverness and saved a day's drive. Proper piers were just being completed, guaranteeing passengers a dry landing; Telford and Mitchell remembered previous occasions when they'd had to mount their horses nearly a quarter of a mile offshore and ride 'mid-leg deep' through the water.

By evening the trusty coachman and landau had transported them about 50 miles. They passed through Tain, whose inhabitants 'entertain a great contempt for Dornoch, the capital of Sutherland, on the opposite side of the bay'; and on to Kincardine where Southey, Telford, and Mitchell put up for the night, leaving the only available rooms at nearby Bonar Bridge to the Rickmans and Miss Emma Piggot.

> Monday, September 6. Walked to Bonar Bridge to join our friends at breakfast. Dornoch Firth, into which the river from Loch Shin discharges itself, runs some 36 miles up the country, and this bridge is 24 from the mouth of the Firth. Upon trying the bottom it was twice pronounced that there was rock; and upon this presumption iron was cast for a bridge of the same proportions as that at Craig Elachie, two such arches being intended. The same moulds were used; but upon further trial it proved that the rock was only on the left bank; and it became necessary to sink 16 feet for a bottom, and besides one iron arch to erect two stone ones … so that the beauty of the structure is destroyed. Yet it is a work of such paramount utility that it is not possible to look at it without delight.

It is clear that Telford originally intended to base his Bonar Bridge on two cast-iron arches; instead he ended up with a mongrel mixture of iron and stone. But even this solution could not last indefinitely: completed in 1812, it survived impacts from fir trunks and a ship, then collapsed in January 1892 when storms overfilled 17-mile long Loch Shin and sent raging floodwater down into Dornoch Firth.

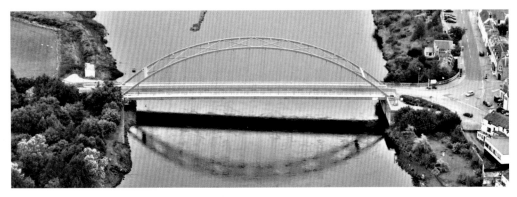

Bonar Bridge (Geograph/Graeme Smith).

The second Bonar Bridge, completed in 1893, had three iron girder spans, of 70, 150 and 140 feet. It was replaced in 1973 by the present structure, a bowstring girder design 339 feet long. Telford would surely have approved such elegant simplicity.

Southey tells a 'remarkable anecdote' about the bridge they saw in 1819:

> An inhabitant of Sutherland, whose father was one of the persons drowned at the Meikle Ferry, over this Firth, in 1809, could never bear to set foot in a ferry boat after that catastrophe, and was thus cut off from communication with the south till this bridge was built. He then set out on a journey. "As I went along the road by the side of the water," said he, "I could see no bridge: at last I came in sight of something like a spider's web in the air – if this be it, thought I, it will never do! But presently I came upon it, and oh, it is the finest thing that ever was made by God or man!"
>
> Mitchell had a most providential escape from that tragedy at the Meikle Ferry. He was pushing his horse to be in time, and was only about three minutes too late. When he came upon the rising ground nearest the shore, the boat had sunk, and of 109 persons, some half dozen only were swimming for the land!

So it seems that Telford's bridge, as well as improving the Highland road network, had helped a small Highland community come to terms with a dreadful tragedy.

A marble tablet at the north end of the bridge had been engraved to honour the commissioners responsible for improving the road network:

> **Traveller, Stop and read with gratitude the names of the**
> **Parliamentary Commissioners appointed in the year 1803 to direct**
> **the making of above 500 miles of roads thro' the Highlands of**
> **Scotland; and of numerous bridges, particularly those at Beauly,**
> **Scuddel, Bonar, Fleet, and Helmsdale connecting those roads.**

Then followed 11 names, including that of 'Sir Wm. Pultenay, [*sic*] Bart' who had acted as patron to Thomas Telford during his time in Shropshire, built Pulteney Bridge in Bath, and been elected as MP for Shrewsbury in several parliaments. Interestingly he had also served as MP for Cromarty in the 1770s. This man of many parts clearly had interests in the north of Scotland, but he died in 1805, seven years before Bonar Bridge was completed. Apparently the marble tablet acknowledged Thomas Telford's role as 'Architect', but the inscription was full of errors, and 'to crown all it is fixed against the Toll House instead of the Bridge'.

From Bonar Bridge their route skirted the north shore of Dornoch Firth. Southey noticed half a dozen seals, the first he had ever seen, on one of the sand banks; and on the road 'inhabitants of the surrounding country who were on their way in their best attire to a preaching at Dornoch, yesterday having been a Sacrament Sunday'. They stopped at the Clashmore Inn, a 'neat house' built by the Countess of Sutherland's family the previous year and decorated with her crest, a 'Cat o' Mountain rampant'. Thence 5 miles north to Little Ferry where Loch Fleet meets the North Sea, with Dunrobin Castle in the distance.

Little Ferry had long been an inconvenience for travellers, mainly because the service was at the mercy of tides and currents. Mitchell had previously experienced delays of up to three hours and must have been keen on an engineering solution. Telford initially proposed building new piers for the ferry at the mouth of Loch Fleet; but William Young, factor to the Sutherland Estates, suggested a permanent embankment and bridge at the other end of the loch, to carry the new parliamentary road on towards Wick and Thurso. This was approved, Telford designed it, and it was built between 1814 and 1816 with local labour. Young wrote that as the last breach was about to be filled 'We are all prepared to battle the sea with 600 men. Walklate is brewing 30 hogsheads of ale and the Baker is getting 40 Boils of Meal converted into Bread, for the people must be fed.' The Countess of Sutherland crossed Fleet Mound on 26th June 1816 'carriage and all'.

The mound's great earthen embankment is extraordinary: nearly 1,000 yards long, it reminds us of Telford's earlier embankment leading onto the Pontcysyllte Aqueduct and the great clay embankment at Clachnaharry on the Caledonian Canal. The bridge that punctuates it is also extraordinary because the arches are fitted with non-return flap valves that allow the river to flow out but stop the sea coming in. At certain times of year the flaps are opened mechanically to allow salmon to swim upriver.

Southey was astonished by Fleet Mound:

> You perceive at once the simplicity, the beauty, and the utility of this great work; but you are not at first fully sensible of its grandeur, the straightness of the line appearing to diminish the length. About 400 acres which were formerly under water, are thus left dry, and grass is already beginning to grow upon what in a few years will become a fertile Carse. Moreover, lands of some extent, which were liable to be flooded, are now secured from that evil. The cost was about 8000£. Lord Gower gave one, the rest was drawn in equal shares as usual from the County and the Public Grant. Lord Gower gains the land which is recovered from the sea; and also derives more advantage from the road, and the dry passage to Dunrobin, than any other person.

View over Fleet Mound. Freshwater River Fleet is on the left,
saltwater Loch Fleet on the right (Geograph/Andrew Tryon).

Sluice gates on the bridge at Fleet Mound (Geograph/Andrew Tryon).

The landau has carried them all the way to Sutherland, and they are now within sight of Dunrobin Castle, the home of the Marquis of Stafford and his wife Elizabeth, Countess of Sutherland. You may recall that Southey had by chance noticed their agent, James Loch, in the inn at Stonehaven 11 days previously, and had commented that Loch was "at present exposed to much unpopularity and censure for the system which he is pursuing. Without knowing the merits of the case, his appearance would prepossess me in his favour". In the meantime Southey had spent many hours in Telford's company, aboard the landau and in hotel rooms, so they must surely have discussed "the merits of the case" together.

Loch's unpopularity and censure arose because he was carrying out the instructions of the marquis and his wife to move large numbers of impoverished tenants off their vast estates, and replace them with sheep. By 1819 the Highland clearances were well under way, and Loch was seen by many as a master planner and operator. A Lowland Scot who had trained as a lawyer, he was content to work from Edinburgh or London and delegate the execution of his plans to subordinates in Sutherland.

A special horror had taken place in 1814, 'The Year of the Burnings', when tenants were torched out of their homes in Strathnaver by the notorious Patrick Sellar; and now, five years on, the burners were back. In June 1819 the Scotsman newspaper reported: 'It is said that a posse of men (with legal warrants be it observed) are parading the county of Sutherland and ejecting poor Highlanders from the homes of their fathers'. The publicity created a stink, and

Dunrobin Castle, near Golspie (Geograph/Andrew Tryon).

Southey may have espied Loch in Stonehaven as he journeyed south after a crisis meeting with his aristocratic employers in Dunrobin Castle.

Now that Southey is, so to speak, within a stone's throw of the lions' den, he feels a need to address the issue properly. His fellow travellers, Telford, Mitchell, and Rickman, have been involved for years in the government's programme of infrastructure improvements in the Highlands, most of which is welcomed by local aristocrats and landowners. Actually Loch was offering the Sutherland Estates a strategy that not only substituted sheep for people but also turned rough tracks into roads, bridged rivers and streams, created new harbours for the fishing boats that were to be crewed by the displaced crofters, and built settlements close to the coast with decent housing in place of primitive huts – in other words, a general redirection and modernisation of the estates that, it was claimed, would greatly benefit the (few) people who remained. Unsurprisingly, critics argued that the aristocrats' motivation was essentially greed, plus a desire to get rid of awkward tenants who could only pay peppercorn rents.

Southey now devotes three pages of his journal to this thorny issue, and the following excerpts give a good idea of his attitude:

> There is at this time a considerable ferment in the country concerning the management of the M. [Marquis] of Stafford's estates: they comprise nearly 2/5ths of the county of Sutherland, and the process of converting them into extensive sheep-farms is being carried on. A political economist has no hesitation concerning the fitness of the end in view, and little scruple as to the means. Leave these bleak regions, he says, for cattle to breed in, and let men move to situations where they can exert themselves and thrive. The traveller who looks only at the outside of things, might easily assent to this reasoning. I have never – not even in Galicia – seen any human habitations so bad as the Highland "black-houses"; by that name the people of the country call them, in distinction from such as are built with stone and lime. The worst of these black houses are the bothies – made of very large turfs, from 4 to 6 feet long, fastened with wooden pins to a rude wooden frame …
>
> Here you have a quiet, thoughtful, contented, religious people, susceptible of improvement, and willing to be improved. To transplant these people from their native mountain glens to the sea coast, and require them to become some cultivators, others fishermen, occupations to which they have never been accustomed – to expect a sudden and total change of habits in the existing generation, instead of gradually producing it in their children; to expel them by process of law from their black-houses, and if they demur in obeying the ejectment, to oust them by setting fire to these combustible tenements – this surely is as little defensible on the score of policy as of morals. And however legal this course of proceeding may be according to the notions of modern legality, certain it is that no such power can be legitimately deduced from the feudal system, for that system made it as much the duty of the Lord to protect his vassals, as of the vassals to serve their Lord …

*Reconstructed blackhouse at the Highland Folk Museum,
near Kingussie (Wikipedia/Kim Traynor).*

> Except in forcing on this violent change, great good arises where a large estate in Scotland is transferred by marriage to an English owner, English capital and ingenuity being employed to improve it; whereas a native Laird would too probably, like an Irish gentleman, have racked his tenants to support a profuse and wasteful expenditure. Thus in the instance of the Marchioness of Stafford's possessions. They are of enormous extent, tho' they produce not more than between 5 and 6000£ a year; and not only is the whole of the receipts expended in improving them, but about an equal sum from the Marquis's English property is annually appropriated to the same purpose, the Marchioness, much to her honour, having this object at heart.

Perhaps his mixed feelings are what we should expect from an admirer of all that Telford was doing in the Highlands; but also a man with a poet's sensibilities who was struggling with the methods being used to force people off land occupied by their families for generations.

But what would Telford himself have thought? He was devoting himself passionately to the improvement of his country's transport infrastructure, including canals, roads, bridges, and harbours. Why should he think any differently about the northern Highlands? It may help to recap a little.

Back in 1801 Telford had been instructed by the government in London to proceed with a survey and report on the subject of Highland communications. It was an opportunity to serve his countrymen which he seized with both hands. He carried out the survey in double-quick time, reporting to a friend that he had travelled 'along the Rainy West through the middle of the tempestuous wilds of Lochaber, on each side of the habitation of the far famed Johnny Groats, around the shores of Cromarty, Inverness and Fort George, and likewise the coast of Murray'. His report was so favourably received that he was ordered to continue his survey the following summer with special reference to road communications, and to investigate the causes of Highland emigration.

Not surprisingly, Telford's second report to the Treasury recommended roads into the western and northern Highlands (including Sutherland and Caithness), new and improved bridges and harbours, and a canal through the Great Glen. It made such a favourable impression that his countrymen honoured him with a Fellowship of the Royal Society of Edinburgh; and in 1803 the government set up a Commission for the Caledonian Canal, and another one for Highland Roads, Bridges, and Harbours. In both cases Rickman was secretary, and Telford engineer.

So Telford had been intimately involved in plans for the Highlands for about 17 years before he met Southey in Edinburgh, where they were 'upon cordial terms in five minutes'. Throughout their tour Southey displays an appreciation – almost a reverence – for Telford's work, and sees him welcomed – almost lionised – in several aristocratic households. They must have discussed the Highland clearances issue together, and from all we know about Telford's personal qualities – his Laughing Tam image as a youth, fairness in dealings with others, lack of interest in money for its own sake and, most recently, kindness towards a fellow traveller with a 'volcano' in his head – it is hard to believe he would have disagreed with Southey's mixed feelings about James Loch and the Sutherland Estates.

And so, back to Fleet Mound, which Southey nicknamed Telford Mound 'in honour of our excellent companion who has left so many durable monuments of his skill in this country'. It was the most northerly point of their planned tour, so the landau was turned and they started back towards Dingwall via the Clashmore Inn, where the horses were rested and the children were fed, and the 'good hostess' provided whisky flavoured with lemon peel and cinnamon. Along the way Southey noticed a social trend:

> It is a proof of increasing decency and civilization that the Highland philibeg, or male-petticoat, is falling into disuse. Upon a soldier or a gentleman it looks well; but with the common people, and especially with boys, it is a filthy, beggarly, indecent garb … introduced by an Englishman … because the men whom he employed wore nothing but the plaid, and when they were at work, were, as to all purposes of decency, naked.

They passed through Bonar Bridge and dined at Kincardine. Next day, 7 September, they reached Dingwall via the recently completed Fearn Road (now the Struie Road, B9176) that bypassed Tain and cut 12 miles off the journey

> the Fearn Road has been pronounced one of the most perfect lines in the Highlands. It is carried 700 feet above the level of Dornoch Firth; nor is there anywhere a finer specimen of roadmaking to be seen, than where it crosses one dingle on one side, and one on the other; the bridges, the walled banks, the steep declivities, and the beautiful turfing on the slope, which is frequently at an angle of 45, and sometimes even more acute, form a noble display of skill and

View from Struie (Fearn) Hill across the upper reaches of Dornoch Firth towards Bonar Bridge (Geograph/Andrew Tryon).

power exerted in the best manner for the most beneficial purpose. The views over the bay are fine. From this high ground the lake above Bonar Bridge is seen, formed by Shin-water and Rappoch-water.

It is almost as though Southey, having admitted the human cost of the Sutherland clearances, is determined to reassert his belief in the essential goodness of Telford's achievements in the Highlands. He also mentions the greatly missed Matthew Davidson, whose widow had placed gooseberries and apricots in the landau just before they left Inverness:

> On the summit is a point which Mr Telford and Mitchell call Davidson's Crag, because when that humourist was met here one day, descending and leading his horse (it was before the road was made, and he was a timorous rider) his knees trembling as much from fear as fatigue, he curst the place and the Crag too, which, he said, had been making faces at him all the way.

Climbing up and over the Fearn summit must have been emotional for all of them as they remembered a colleague who had worked himself to an early death delivering the commissioners' plans.

West coast excursion

See map on page 69

Some shuffling of horses and wheeled transport was necessary before the men could set out on a planned three-day excursion:

> Wednesday, September 8. Left the Ladies and the children at Dingwall; they were to return in the Coach to Inverness and there wait for us. We with a chaise and Mitchell's gig set off at six, to cross the island by the Loch Carron Road, from sea to sea. The chaise horses had been sent off yesterday, one stage to Garve, and we took a pair of the Coach horses so far, which enabled us to perform the whole journey in one day. North of Inverness, post horses are not to be obtained.

Southey fails to say why they undertook a journey 'from sea to sea' – an obvious distraction from their main anticlockwise route. But as Dingwall, at the head of Cromarty Firth, is a mere 48 miles from the head of Loch Carron, a sea loch opposite the Isle of Skye, they presumably saw it as an opportunity not to be missed – Southey would experience some proper Scottish mountains for the first time since leaving the Trossachs, Telford and Mitchell could inspect a major road under construction, and Rickman would go wherever they and the horses led him.

So they started west from Dingwall, passed through Strathpeffer to Contin, and breakfasted in Garve, a distance of 16 miles. Then into Strath Bran, which leads up through increasingly isolated country to tiny Achanalt:

> The only man whom we saw in a philibeg during this day's journey was a poor idiot, who ran after the chaise, not to beg, but with an idiotic delight at seeing it. The road lies sometimes near, and sometimes along a chain of small lakes, or broads, as some of them might properly be called. The workmen were finishing this division of the road, under the inspection of Mr Christie, the contractor: their tents, which had been purchased from the military stores, were pitched by the wayside, and they had made a hut with boughs for their kitchen – more picturesque accompaniments to so wild a scene could not have been devised. In a country like this, where there is little use of wheel carriages, the road is constructed wholly of gravel, and all the stones are picked out and thrown aside. We went into the inn at Auchnault, a miserable place, bad as a Gallician [sic] posada, or an estallagem [sic] in Algarve. But we tasted whiskey here, which was pronounced to be of the very best and purest, "unexcised by Kings"; and we drank a little milk, on the excellence of which these highlanders pride themselves. The house, wretched as it was, was not without some symptoms of improvement …

They came next to Achnasheen, where the inn was more encouraging: tidier inside, with an open bedstead (a type on which Mitchell had often slept during his travels), and above all an excellent meal for a shilling a head, laced with a dram of Highland hooch.

A train leaving Achnashellach Station (Geograph/David Gruar).

Beyond Achnasheen they entered Glen Carron, passed Loch Gowan and Loch Sgamhain on the left, crossed the watershed, and began a long, slow descent to Achnashellach, Loch Carron, and the western sea.

Achanalt, Achnasheen, Achnashellach: impressive names for diminutive settlements. The first time I came this way was by rail from London, in need of a few days' Highland solitude after an exceptionally busy spell in the capital. Ordnance Survey maps suggested that Achnashellach Station and youth hostel would supply all I wanted, so I packed a small rucksack, donned walking boots, and left Kings Cross that evening. Eagerly anticipating what lay ahead, I changed trains at Inverness the next morning and asked the driver to let me off at Achnashellach, a request stop on a single-track line. Nobody else left the train, nobody got on, and I found myself alone on a diminutive flower-bedecked platform as the train pressed on towards Loch Carron and Kyle of Lochalsh. But I need not have worried: a beckoning wave from the signal box encouraged me to climb the wooden stairs for a chat; and I discovered the strength needed to work old semaphore signals, the way to the youth hostel, and the friendliness of a Highlander contented with a solitary walk of life.

Loch Carron offered Southey his first experience of a western sea loch – and what he assumed to be typical west coast weather:

> The evening set in with rain – which was to be expected in this rainy region. We saw seals swimming in the salt water, and finally on the shore of this long inlet of the sea, we took up our night's abode at Jean-town. Our sitting room is larger than seems either needful or comfortable in such a situation, and there is no air of neatness about it. We had, however, a good meal at tea, excellent butter, barley cake and biscuit (no wheaten bread) and herrings, much smaller than those at Cullen, but delicious enough to vie with them. They are mostly without roes, whence I suppose them to be young fish.
>
> Thursday, September 9. Jean-town, the capital of Applecross's country, and the largest place in the west of Ross-shire, is a straggling but populous village, chiefly or wholly inhabited by fishermen … Great part of the year the men,

Loch Dughaill, close to Achnashellach Station (Paul A. Lynn).

from the nature of their calling, have nothing to do; yet they buy their nets at Inverness, instead of employing some of their leisure hours in making them.

Even if you know the area well, you may be as mystified as I was by the name Jean-town. It turns out to be the original name for Lochcarron village, on the lochside 5 miles from Strome Ferry; but I have no idea who Jean was.

After breakfast they set off for the ferry 'with the intention of crossing there, if Applecross's new boat should be ready, and returning by the Kintail, Glenshiell [*sic*], and Glenmoriston roads'. The plan was to return to Inverness by an alternative, more southerly, route.

But it was not to be, and for reasons that tell us a lot about travel in the West Highlands in 1819. For a start, they were told that the inn known as Shiel House, built at Glen Shiel by the government to offer a night's rest to travellers and their horses, had been commandeered by a certain Mr Dick, who had shut it down, behaved with 'great insolence' to Mitchell, and fallen out with the commissioners. This meant that the 70 miles from the far side of Loch Carron to Inverness would have to be tackled without any certainty of a decent overnight stop en route. Even so, they trusted to luck and arrived at the ferry hoping to find the new boat, faithfully promised by Applecross, ready and willing to take them across. But it was not there. What was there – presumably a rowing boat plus oarsman – was useless for horses and carriages.

Loch Carron (Geograph/Richard Dorrell).

However, instead of turning back immediately, annoyed and disconsolate, they decided to take the ferry as foot passengers and investigate the far side of the loch:

> Loch Carron is a beautiful inlet … inclosed by mountains on three sides, and on the fourth the mountains in the Isle of Skye are seen at no great distance. Ours was the first carriage which had ever reached the ferry, and the road on the southern shore, up which we walked, had never yet been travelled by one. We went up the hill so as to command the descent along which it inclines toward Loch Alsh – a district, not a lake – and communicates by Kyle Haken [Kyleakin] Ferry with the Isle of Skye, where an hundred miles of roads have been made by the Commissioners. To hear of such roads in such a country, and to find them in the wild west Highlands is so surprizing, everything else being in so rude a state, that their utility, or at least their necessity, might be doubted, if half the expense were not raised by voluntary taxation.

But it *was* raised by taxation, and the local lairds who contributed gained handsomely from the scheme, especially as many of their tenants owed large amounts of rent and were willing to discharge their debts by labouring. If the estimated cost of a road was £5,000, the government paid the lairds £2,500, and the tenants contributed £5,000 worth of labour. 'Thus they [the lairds] were clear gainers by all which they received, and by the improved value of their estates.'

Telford and Mitchell must have been doubly disappointed by the ferry fiasco – not only by having to return the way they had come, but also by missing the chance of casting their expert eyes on 100 miles of new roads on Skye; and Southey, Lakeland man and lover of mountain scenery, would surely have marvelled at the Cuillins. But they had no choice but to backtrack towards Jean-town:

> The weather had been stormy when we crossed, and it rained heavily while we were on the way to Craigie House, beyond which we could not proceed this day, for want of decent accommodation at night. A girl past us in a cart carrying an umbrella … Craigie House is smaller than the inn at Jean-Town. They have only two beds for strangers, but will make up a third on six chairs, by robbing one of the others …
>
> Friday, September 10. – The upper story of Craigie House is constructed in the roof so incommodiously that a large corner is, of necessity, cut from the doors; and a man must beware of his head, unless he walks in the middle of the room. I could neither get in, nor out of my chair-bed, nor sit upright in it, without management. The chairs were lengthened by placing a chest at the bottom of the bed; but the chest and the chairs were not upon the same level, so that my feet had a step to go down. However I slept well; and the shifts which I was fain to use in rising and dressing, there being no passage between the two beds, were matter of merriment. Quilts of ornamental patchwork, as at Auchnault.

It was a beautiful morning with mountains, valleys, and streams dressed in sunshine. They passed through Achnashellach and Achnasheen before breakfasting at Achanalt on good pink potatoes (no bread) and a cold sheep's head, which Southey thought very good 'because of the skin, and the flavour which had been given it by singeing'. Then on to Strath Garve, where he noticed a strange type of dry stone walling so wretchedly made that it would be destoyed by 'the slightest push, or the first storm of wind'. But Telford, brought up in the Borders, disagreed, explaining that sheep, seeing light through gaps between the stones, would never attempt to jump the wall and bring it down with their heels. He called it a Galloway Dike.

They reached Dingwall in time for a late dinner; and next morning crossed the Black Isle to take Kessock Ferry, 'the best in Scotland', which crossed the narrows between the Moray and Beauly Firths and landed them on the outskirts of Inverness. However

> the best ferry is a bad thing. They have no good means of getting carriages on board, and there was considerable difficulty with one of the horses. As soon as we arrived at our Inn, I sent for Mr Kennedy. He was surprized to see how compleatly his ointment had done its work. The proud-flesh is gone, and the tumour has nearly, and he says it will soon heal.

So ended their three-day excursion, a mixture of achievement and frustration. And Southey's volcano had benefited, probably from the west coast sea air.

Kessock Ferry in 1971. It was withdrawn in 1982 when Kessock Bridge was opened (Geograph/Gordon Spicer).

Down the Great Glen to Fort William

See map on page 69

> Sunday, September 12. Walked to the mouth of the [Caledonian] Canal. It opens into a fine road-stead doubly sheltered by the opposite coast of the Black Isle; and by the points of Fort George and Channerty Point, which cover the entrance of the bay. The masonry at the mouth is about ten feet above high water mark: the locks large enough to admit a 32 gun frigate, the largest which has ever been made. There was a difficulty at the mouth from the nature of the bottom, being a mud so soft that it was pierced with an iron rod to the depth of sixty feet. A foundation was made by compressing it with an enormous weight of stones, which were left during twelve months to settle, after which a pit was sunk in it, and the [Clachnaharry] sea lock therein founded and built. This was a conception of Telford's, and had it not been for this bold thought the design of the canal must have been abandoned. The length of the basin is 800 yards, the breadth 150. Already the sea has, as it were, adopted the outworks, and clothed the embankment and the walls with sea-weed.

Robert Southey and company are back in Inverness, and the Caledonian Canal, 12 years into build and 3 years from completion, is about to give the poet his most satisfying insights into the engineering genius of Thomas Telford. We know this because he devotes 44 pages of his journal to the canal and its surroundings as they travel down the Great Glen to Fort William; and because many commentators and writers (including Tom Rolt) have applauded the poet's eyewitness account of work in progress – the impressions of a highly intelligent and observant onlooker who was not a civil engineer.

They boarded a boat at Muirtown and experienced their first locks, 'shut in between such tremendous gates on two sides, and such walls of perpendicular masonry on the other

Left: Inverness, with the 1800 Greig Street pedestrian bridge across the Ness (Geograph/Rob Farrow).

Right: Muirtown Locks, Inverness (Geograph/Stephen McKay).

two, the situation might have afforded a hint for a Giant's dungeon'. Beyond the locks they sailed along Loch Dochfour as far as its junction with Loch Ness, where waves built up along 22 miles of water by a strong southwesterly had thrown up a high beach of pebbles, ridge upon ridge, and were breaking 'with the voice of an ocean'.

> You see along the whole expanse; the mountain sides, on either hand, confine the view … terminated by the mountains beyond Fort Augustus … I have never seen so large a lake as Loch Ness, which could be seen as a whole in one unbroken line … more impressive than that of Neufchatel … Loch Ness has been carefully sounded, and found to be 129 fathoms deep … how can this prodigious hollow have been formed? It never freezes … The water is said to be unwholesome, acting as a purgative on man and beast.

Looking south along Loch Ness (public domain).

A Wade bridge on the military road near Foyers (Geograph/ John MacKenzie).

They spent the next day in Inverness, sightseeing, writing letters and in Southey's case updating his journal. Wednesday 15 September saw them starting on a road journey right down the Great Glen – Rickman and Mitchell along the west shore of Loch Ness by the Glen Morriston road; Telford, Southey, and the rest along the east shore by Wade's military road:

> There is more of our Lake-land character upon this road than in any other part of Scotland thro' which my way has lain – rocks, fern, and heather upon the side of the green hills, the lake below, and on the opposite mountain, where the rain has laid it bare in streaks, there is the same red colouring as at Buttermere and Wasdale.

They visited the remains of General Wade's hut near the village of Foyers, built of mud and straw on wooden framing, and Southey laments that mischievous travellers had picked it to bits, destroying the interior 'colour and gloss' caused by peat smoke, which must have made it 'handsome as well as peculiar'. Southey puts in a good word for the smoke, which he finds 'clean and agreeable', but admits it damages the sight, as proven by 'the blear eyes which are here so common among old people'. Unfortunately a visitors' book formerly kept at the Foyers Inn, in which travellers from General Wade's time onwards had signed their names and expressed feelings personal and political, had been stolen by 'some scoundrel' a few years previously; but the inn offered the travellers a welcome meal of potatoes, butter, and milk.

Next on the agenda was the Fall of Foyers, a famous waterful which in full flood 'might perhaps bear comparison with any single fall of the Reichenbach'. They reached it courtesy of Wade's road, but Southey is again on the warpath – not about the general's bridges, nor the roads themselves, but the routes he had chosen for them. These often followed old horse tracks and crossed hills 'with great difficulty and labour … this is neither agreeable to horses, nor drivers, nor nervous travellers'. Southey notes that between the General's Hut and Fort Augustus the road climbed to about 1,500 feet above Loch Ness, 'half the height of Skiddaw'.

Continuing their journey, with Loch Ness very much on his mind, Southey recalls a near-calamity:

> Loch Ness was violently agitated at the time of the great earthquake of 1755, which destroyed so large a part of Lisbon. The waters of the Lake were driven up the [Loch] Oich more than two hundred yards, with a head like a bore or hygre, and breaking on its banks in a wave about three feet high. Thus it continued to ebb and flow for more than an hour. About eleven o'clock it drove up the river with greater force and overflowed the bank to the extent of thirty feet. A boat near the General's Hut was three times dashed on shore, and twice carried back; it filled with water, and its load of timber was thrown ashore. But no agitation was felt on the land. Such is the account given in the Survey of Moray. Some future Humboldt will avail himself of it in the first geological map which shall lay down the course of earthquakes.

A stretch of Wade's road between Inverness and Fort Augustus (Geograph/David Ayrton).

Apparently Telford had written from Inverness to the landlord at Fort Augustus, asking if he could reserve rooms, 'no unnecessary precaution in these touring times, especially for so large a party'. But the Post Office failed to deliver the letter, as had happened at Jean-town the previous week – 'just another instance of negligence in the management of the post'. On arrival Southey met Marshall the landlord, dubbed 'the Field Marshall' by Telford, and decided that had he been six inches shorter he would have been as broad as long, a convincing 'Knight Errant's Dwarf'. More importantly for their comfort, the inn was 'inconveniently built, and not well situated' – far inferior to the house erected nearby for Mr Cargill, a Newcastle man and master mason of the stupendous canal works in progress at Fort Augustus.

There was enough daylight before dinner for them to witness a dredging machine at work – 'an engine of tremendous power, bringing up its chain of buckets full of stones and gravel, or whatever comes in its way'. It was Southey's first sight of heavy canal engineering, and the following days would reveal a great deal more. In the meantime he penned some thoughts about the building which gave Fort Augustus its name:

> the Fort itself is very pretty – a quiet collegiate sort of place, just fit for a University, if one were to be established, or for a Beguinage [convent of the Beguine Sisterhood], if the times and the situation served. The guns have lately been removed … However it was military enough for its purpose, and proved an effectual check upon the wild and disaffected clans whom it was meant to curb … An officer fond either of country sports, or of reading, or of quiet life and a picturesque country, would think himself well off in such quarters. But such officers are not the sort of men which our army hitherto has usually bred: and the people who have been stationed here, have nothing to do in their profession, and being incapable of doing anything out of it, have always been engaged in petty disputes, idleness and ennui generating peevishness, discontent and ill will. It is said that the expense of sending persons from Edinburgh to examine into the mutual accusations of these poor creatures, has frequently amounted to more in the year than the whole regular cost of the garrison.

Where he got all this from is unclear. Maybe he learned it en route, and inserted it in his journal later. It confirms him as an unlikely commentator on military matters, a poet somewhat out of his depth: no shots fired, magnificent weeds at Fort George on 3 September; guns removed, petty disputes at Fort Augustus on 15 September. Time to move on.

Loch Ness, which had contributed 22 miles to Telford's canal without requiring any engineering, was now behind them. The next few miles were entirely different, and would stimulate Southey's imagination enormously – starting with the five locks at Fort Augustus which had been giving Telford a huge headache:

> Thursday, September 16. Went before breakfast to look at the Locks, five together, of which three are finished, the fourth about half-built, the fifth not quite excavated. Such an extent of masonry, upon such a scale, I have never before

beheld, each of these Locks being 180 feet in length. It was a most impressive and rememberable scene. Men, horses, and machines at work; digging, walling, and puddling going on, men wheeling barrows, horses drawing stones along the railways. The great steam engine was at rest, having done its work. It threw out 160 hogsheads per minute; and two smaller engines (large ones they would have been considered anywhere else) were also needed while the excavation of the lower docks was going on; for they dug 24 feet below the surface of water in the river, and the water filtered through open gravel. The dredging machine was in action, revolving round and round, and bringing up at every turn matter which had never before been brought to the air and light. Its chimney poured forth volumes of black smoke, which there was no annoyance in beholding, because there was room enough for it in this wide clear atmosphere. The iron for a pair of Lock-gates was lying on the ground, having just arrived from Derbyshire: the same vessel in which it was shipt at Gainsborough, landed it here at Fort Augustus. To one like myself not practically conversant with machinery, it seemed curious to hear Mr Telford talk of the propriety of weighing these enormous pieces (several of which were four tons weight) and to hear Cargill reply that it was easily done.

This lengthy paragraph is Southey's first attempt to describe the extraordinary scene that the construction of canals – like railways 20 years later – could impose on a normally quiet community. And a great job he makes of it: massive masonry, human bustle, huge lock gates from Derbyshire, a dredging machine belching black smoke. The organisation required must have been meticulous – and certainly greater than that needed to deliver a letter from Fort

Left: The channel between Loch Ness and the flight of locks at Fort Augustus. Lock no.5, which gave Telford enormous trouble, is just off the photo to the right (Geograph/Jennifer Jones).

Right: A Fort Augustus lock by night (geograph/Oliver Dixon).

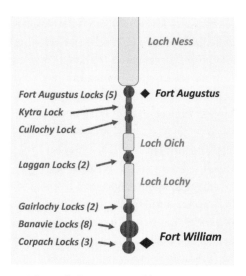

Fort Augustus Locks (5)

Kytra Lock

Cullochy Lock

Laggan Locks (2)

Gairlochy Locks (2)

Banavie Locks (8)

Corpach Locks (3)

Loch Ness

◆ Fort Augustus

Loch Oich

Loch Lochy

◆ **Fort William**

The Caledonian Canal between Fort Augustus and Fort William.

Augustus's post office to its rather shoddy inn. The villagers, used to a quiet life since the fort had lost its defensive purpose, must have been bewildered.

What Southey could not have known was that Lock no.5 at Fort Augustus, which led out into Loch Ness, would continue to give Telford trouble and delay the opening of the whole canal – a veritable nightmare.

Breakfast that morning was full of chatter about what they had just seen – except when Tom, the Field Marshall's pet ram, scared and distracted the reassembled party, and especially the children. Tom was a large, powerful beast who, when offered a piece of bread soon 'pushed on for more with a strength which it was not easy to control, and endangered the breakfast table'.

At this point you may like to refer to the accompanying map of the Caledonian Canal from Fort Augustus to Fort William, via Lochs Oich and Lochy. In the next few days the party would cover all this ground, and Southey would complete his canal-building education.

After breakfast they set off towards Loch Oich, the summit level of the canal. Southey had been primed to expect 'the greatest work of its kind that has ever been undertaken in ancient or modern times, in all stages of its progress – directed everywhere by perfect skill, and with no want of means'. He was not disappointed:

> The Oich has, like the Ness, been turned out of its course to make way for the Canal. About two miles from Fort Augustus is Hytra [Kytra] Lock, built upon the only piece of rock which has been found in this part of the cutting – and that piece just long enough for its purpose, and no longer. Unless rock is found for the foundation of a lock, an inverted arch of masonry must be formed, at very great expence, which after all is less secure than a natural bottom. At this [the northern] end of Loch Oich a dredging machine is employed, and brings up 800 tons a day. My Hughes, who contracts for the digging and deepening, has made great improvements in this machine. We went on board, and saw the works; but I did not remain long below in a place where the temperature was higher than that of a hot house, and where machinery was moving up and down with tremendous force, some of it in boiling water.

Southey has a happy knack of enlivening his account with an eclectic mixture of anecdotes and observations unconnected with the work in progress. He informs us that there was a wooden bridge over the River Oich, from which a post chaise with two people had recently plunged into the river; a mountain visible to the northwest, known by the 'vile name' of Glengarry's Bowling Green, was the finest mountain he had yet seen in Scotland; the highest level reached by the canal was less than the height of Oxford Street above the River Thames; and it was an especially beautiful day, 'when light showers were flying about, and there was a lovely skyscape of soft and silvery clouds in the West'.

They next took a walk to view the massive earthworks between Lochs Oich and Lochy:

> Here the excavations are what they call "at deep cutting", this being the highest ground in the line, the Oich flowing to the East, the Lochy to the Western sea. This part is performed under contract by Mr Wilson, a Cumberland man from Dalston, under the superintendence of Mr Easton, the resident Engineer. And here also a lock is building. The earth is removed by horses walking along the bench of the Canal, and drawing the laden cartlets up one inclined plane, while the emptied ones, which are connected with them by a chain passing over pullies, are let down another. This was going on in numberless places, and such a mass of earth had been thrown up on both sides along the whole line, that the men appeared in the proportion of emmets to an ant-hill, amid their own work. The hour of rest for men and horses is announced by blowing a horn; and so well have the horses learnt to measure time by their own exertions and sense of fatigue, that if the signal be delayed five minutes, they stop of their own accord, without it … The workmen are mostly steady industrious men, who work by the piece, and with a good will, because they are regularly paid. They have communicated some industry to the inhabitants. We saw large fields of potatoes, intended for their consumption.

Southey is referring to the stupendous Laggan Cutting, described by Tom Rolt as a work that 'dwarfs in its scale almost all the works of the later railway builders which the world was soon to marvel at'.

A calm day at Loch Oich (Geograph/Richard Webb).

At the end of a very busy day they returned to their 'dirty quarters' at Fort Augustus. The following day was comparatively leisurely, Telford and Rickman attending to business matters and Southey again updating his journal.

> Saturday, September 18. Our comical host every day exhibited at breakfast the fish which he intended for our dinner, and explained the difference between the river and the lake trout. The latter were much darker, and neither kind so much spotted as our English trout. I observed also that the flesh of neither (if I may speak of the flesh of fish) was so red when they came to table. As for their comparative goodness … I can say nothing, for they were cut across, like crimped fish, and then broiled, till the flavour, whatever it might have been, was broiled out of them. Moreover, melted butter, which the Scotch use with nothing, except fish, is in Scotland such a vile mixture of flour and butter, that it is not fit to be used with anything … Dr Johnson says in one of his letters "the best night I have had these twenty years was at Fort Augustus". He therefore remembered the place with pleasure. And so shall I – always excepting the quarters, which could not have been filthier in his time than they are at present.

How Southey recalled the place with pleasure is hard to imagine. Perhaps the passage of time had softened its worst aspects and allowed the extraordinary characters of the Field Marshall and Tom the sheep to shine through the murk – an experience possibly earmarked for future use by the poet laureate.

In any case Southey, by the time he regained Fort Augustus, had become mesmerised by Telford's stupendous works of engineering, a sublime meeting of science and art. He was already storing it all up, and in 1822, the year in which the Caledonian Canal was opened, he produced three 'Inscriptions for the Caledonian Canal' for presentation to Telford. One was titled 'At Fort Augustus' (the word 'glede' in its opening line probably refers to a red kite):

> Thou who hast reached this level, where the glede,
> Wheeling between the mountains in mid-air,
> Eastward or westward as his gyre inclines,
> Descries the German or the Atlantic Sea,
> Pause here; and, as thou seest the ship pursue
> Her easy way serene, call thou to mind
> By what exertions of victorious art
> The way was opened. Fourteen times upheaved,
> The vessel hath ascended, since she changed
> The salt sea water for the highland lymph;
> As oft in imperceptible descent
> Must, step by step, be lowered, before she woo
> The ocean breeze again. Thou hast beheld
> What basins, most capacious of their kind,

Enclose her, while the obedient element
Lifts or depones its burthen. Thou hast seen
The torrent hurrying from its native hills
Pass underneath the broad canal inhumed,
Then issue harmless thence; the rivulet
Admitted by its intake peaceably,
Forthwith by gentle overfall discharged:
And haply too thou hast observed the herds
Frequent their vaulted path, unconscious they
That the wide waters on the long low arch
Above them, lie sustained. What other works
Science, audacious in emprize, hath wrought,
Meet not the eye, but well may fill the mind.
Not from the bowels of the land alone,
From lake and stream hath their diluvial wreck
Been scooped to form this navigable way;
Huge rivers were controlled, or from their course
Shouldered aside; and at the eastern mouth,
Where the salt ooze denied a resting place,
There were the deep foundations laid, by weight
On weight immersed, and pile on pile down-driven,
Till steadfast as the everlasting rocks
The massive outwork stands. Contemplate now
What days and nights of thought, what years of toil,
What inexhaustive springs of public wealth
The vast design required; the immediate good,
The future benefit progressive still;
And thou wilt pay the tribute of due praise
To those whose counsels, whose decrees, whose care,
For after ages formed the generous work.

On their way to Fort William they stopped at Invergarry to meet the local chieftain, Alexander MacDonell of Glengarry, who had issued an invitation to Rickman and Telford and received them 'with much civility and satisfaction'. He regretted that they had not been able to attend his 'great entertainment' the day before, put on for locals in their traditional Highland dress – a deer hunt at which one animal was shot by Glengarry, another by his chief huntsman, and a third seized by dogs 'of the greyhound make, but neither quite so large, nor quite so slender; and most of them shaggy – as wild in appearance as the mountains on which they pursued their prey'. During their visit Glengarry's son, a lad of about 14, appeared in 'the dress' with a hair cartridge-box hanging in front of him, and a belt containing a sheath with knife, fork, and dirk around his waist. This led to a conversation about bloodthirsty weapons of various shapes and sizes, which probably left Telford and Rickman unmoved and Southey

*Left: Invergarry Old Castle, former stronghold of the
Clan MacDonell of Invergarry (Geograph/Astrid H).*

*Right: High Bridge over the River Spean was already in a perilous state
when Southey crossed it in 1819 (Geograph/Steven Brown).*

disconcerted. The contrast between Glengarry in his pomp and the Field Marshall of Fort Augustus in his decrepit inn could hardly have been greater.

Southey presumably knew that their double-faced host had previously proved highly awkward to Telford and Rickman over land needed for the Caledonian Canal, and that he was Walter Scott's model for a haughty and flamboyant Highland chieftain in the historical novel *Waverley*, published in 1814. During the controversial visit of King George IV to Scotland in 1822 Glengarry, whose clan had consistently supported the Jacobite cause, had arrogantly made unauthorised appearances that exasperated Scott and the other organisers. Unfortunately for his long-term reputation, he also contributed to the Highland clearances by substituting sheep for people, causing many of his clansmen to emigrate to Canada. On many counts, Glengarry was a force to be reckoned with.

Their road now ran alongside Loch Lochy. Letter-Finlay, about halfway along the loch, provided an inn to rival Fort Augustus for dirt and discomfort, but redeemed itself with a good fire, biscuits, cheese, milk, and whisky. The military road, maintained by the commissioners, soon left the side of the loch and headed inland over wild country, crossing the River Spean

by what is properly called High Bridge: the bridge is in a perilous state, and it will be well if it stands till the Commissioners effect their object of turning this road to join the Laggan road, where a new bridge has been erected over the same river. We soon came in sight of Ben Nevis, a precipitous, rugged, stony, uninviting mountain, looking as if it had been riven from the summit to the base, and half of it torn away. It is an aweful mass, and may well be called Big Ben – yet not the greatest of all Bens; for the Ben of Bens is Ben Jonson.

Fort William provided lodgings for the next four nights. Their first excursion was to Corpach at the southern end of the canal, but to reach it they had to cross the River Lochy – 'there are no piers, and we were carried to and from the boat on men's shoulders'. Corpach sea lock was full and a sloop was lying in the basin. They followed the canal bank as far as Neptune's Staircase:

> Six [of the eight locks] were full and overflowing; and when we drew near enough to see persons walking over the lock-gates, it had more the effect of a scene in a pantomime, than of anything in real life. The rise from lock to lock is eight feet, 64 therefore in all; the length of the locks, including the gates and abutments at both ends, 500 yards – the greatest piece of such masonry in the world, and the greatest work of its kind, beyond all comparison.
>
> A panorama painted from this place would include the highest mountain in Great Britain, and its greatest work of art … The Pyramids would appear insignificant in such a situation, for in them we should perceive only a vain attempt to vie with greater things. But here we see the power of nature brought to act upon a great scale, in subservience to the purposes of man: one river created, another (and that a huge mountain stream) shouldered out of its place, and art and order assuming a character of sublimity.

Southey took a great interest in other canal features, including culverts, minor aqueducts, and sluices – especially the Loy Sluices on the way to Gairlochy 'by which the whole canal from the Staircase to the Regulating Lock (about six miles) can be lowered a foot in an hour'. He is amazed that 'three small sluices, each only four feet by three', could produce an effect to rival mighty Swiss waterfalls, and wonders what the Bourbons would have given for such a cascade at Aranjuez or Versailles. 'But the prodigious velocity with which the water is forced out by the pressure above explains the apparent wonder.' Southey is in danger of becoming a civil engineer himself.

They passed Gairlochy and continued for a mile or so on the north side of Loch Lochy – Mitchell on horseback, Telford and Rickman in Mitchell's gig, Wilson and Southey in Wilson's – then branched off to Achnacarry, the home of Cameron of Lochiel near the head of Loch Arkaig. The 18-mile loch impressed Southey more than any other he had seen, apart from Loch Katrine in the Trossachs. Unfortunately, however,

> the owner of this lake, and of the whole beautiful country round, is a poor creature wholly unworthy of his fortune. The family estate when it was restored produced 500£ a year; it now produces 7000£. 3000£ are settled upon his wife who lives about Edinburgh, separate from him. The estate is in the hands of trustees, and he lives miserably in London upon 600£ a year, kept needy by his debauched course of life, and eking out this pittance by cutting down his woods! The roads at this time are almost destroyed by the carriage of his timber.

It seems clear that the 'poor creature' was Cameron of Lochiel, but Southey does not say so, nor whether they had been invited to call in at Achnacarry. But this seems likely – why else would they have diverted away from the canal? Telford probably expected Lochiel to discuss maintenance of the road from Gairlochy to Achnacarry, or improvements to his estate; but it seems the party arrived only to learn that he was in London.

Whatever the details, Southey is thoroughly stirred up. He vents his anger about the 'restoration' of Lochiel's estate, the behaviour of its current owners, and ongoing social problems in the Highlands:

> The restoration of the forfeited estates has produced no good in the Highlands. As an act of grace it carried with it not the appearance only, but the reality of great injustice, in restoring those families who were implicated in the rebellion of 1745, and not the sufferers of 1715, who had surely more claim to indulgence. Far better would it have been for the country in general, and especially for the poor Highlanders, if the estates had been retained as Crown lands, and leased accordingly, or even sold to strangers. The Highland Laird partakes much more of the Irish character than I had ever been taught to suppose. He has the same profusion, the same recklessness, the same rapacity; but he has more power, and he uses it worse; and his sin is the greater, because he has to deal with a sober, moral, well-disposed people, who if they were treated with common kindness, or even common justice, would be ready to lay down their lives in his service … Some fifty land-Leviathans may be said to possess the Highlands … A few of these are desirous of improving their own estates by bettering the condition of their tenants. But the greater number are fools at heart, with neither understanding nor virtue, nor good nature to form such a wish. Their object is to increase their revenue, and they care not by which means this is accomplished. If a man improve his farm, instead of encouraging him, they invite others to out-bid him in the rent; or they dispeople whole tracks to convert them into sheep-farms.

And so he goes on – for another page and a half. It is as though the radical, youthful Southey, who had once worshipped the French Revolution and dreamed of creating an idealistic community on the banks of the Susquehanna River in America, has been reborn. It is by far the most outspoken of several journal entries berating Scottish aristocrats for their behaviour. And he goes on to suggest ways in which the Lochiels and other lairds could improve the lot of their tenants, including fairer leases and encouragement to bring more land into cultivation – not of grain, but of pasture, potatoes, green crops and even 'sour krout'. With better animal husbandry, including wool spinning and knitting 'like the Welsh', and dairying 'when they have learnt cleanliness', the Highland glens could be as productive as English valleys within half a century.

Historical records confirm what you may already suspect: that the Camerons of Lochiel (like the MacDonells of Invergarry) had been on the wrong side in the two Jacobite rebellions. In 1715 their clan chief escaped into exile in France; in 1745 his successor joined

Achnacarry Castle (Wikipedia/Keeshu).

Bonnie Prince Charlie, and experienced defeat the following year at Culloden. The original Achnacarry House, a wooden building, was burned down by government troops, and the estate confiscated by the Crown. In 1802 Donald Cameron, the 22nd clan chief, started to rebuild the house as a castle in Scottish Baronial style; but it was, to put it mildly, a hit-and-miss project punctuated by long periods of inactivity. Eventually he became totally disillusioned with the place, left it, and never returned. In 1837 his successor completed it as 'a handsome residence worthy of the chief', but this was 18 years after Southey's outburst.

Our travellers agreed on a rearrangement of transport; they returned to Fort William, with Southey and Rickman suffering in Mitchell's gig:

> Had the distance been a few miles farther, I believe neither my poor pantaloons, not my poorer flesh, nor the solid bones beneath, could have withstood the infernal jolting of this vehicle, tho' on roads as smooth as a bowling green. As for M [Mitchell], he is so case-hardened that if his horse's hide and his own were tanned, it may be doubted which would make the thickest and toughest leather.

The poet's equilibrium, disturbed by Achnacarry, was finally upset by a rough ride in a rotten gig. The morning's wonderful visit to Corpach, Neptune's Staircase, and the Loy Sluices must have seemed a distant memory as he 'returned to dinner so late, that before we rose from table it was nearly ten o'clock'.

Next morning the men began another expedition of their own, leaving the ladies and children to pass the time as best they could in Fort William. The men's destination was the Parallel Roads of Glen Roy, a celebrated set of horizontal terraces high up on the steep sides of a valley that winds into the mountains from Roybridge, about 12 miles northeast of Fort

William. Ancient folklore held that the 'roads' had been constructed by humans, but by 1819 it was generally agreed that they must be natural in origin, even though there was at the time no satisfactory explanation.

Telford, Rickman, and Southey reached Roybridge in the landau, Mitchell and Wilson on horseback. All five then rode into remote Glen Roy, with Wilson as their guide:

> In no part of our journey could fine weather have been more desirable, and never was there a finer day. Glen Roy is the loveliest glen which I have seen in Scotland; it is very narrow, beautifully green, and has a clear, sparkling stream, and the ascent for 100 or 150 yards is thickly sprinkled with alders growing, not like bushes, but like trees in an orchard. The Parallel Roads are among the most extraordinary objects in Europe or in the world. How would they have excited the astonishment and the speculations of learned men if they had been discovered in Asia or America! Humboldt would have travelled to the Antipodes to see them. They are three broad and distinct terraces strongly defined, upon both sides of the glen, and of those glens which communicate with it; all at a great height, perfectly level, and perfectly parallel with each other, extending as far as we could see, which was several miles, and comprising in all not less than an hundred miles. It would not have been possible to visit this extraordinary scene with better companions, than such a surveyor as Mr Telford, such a practical workman as Wilson, and such a clear, quick, accurate scrutinizer as R [Rickman], the strongest-headed and most sagacious man whom I have ever known.

The Parallel Roads of Glen Roy, highlighted by recent snowfall (Wikipedia/Richard Crowest).

So sagacious, indeed, that Southey goes on to fill two pages of his journal with Rickman's observations on the 'roads', especially their remarkable horizontality and precise vertical separation from one another. He concludes:

> The theory which attempts to account for these roads by the action of water had never for a moment appeared tenable, and here upon the spot its absurdity was at once seen and demonstrated: for, as Mr Telford pointed out, it was manifest that the roads had been made since the ravines – to use no farther argument …
>
> For what can these surprizing roads have been made? A genuine Ossian would probably have informed me. The only information which can now be hoped for, must be sought from etymology. It is likely that the hills and glens retain in their names some allusion to the actions of which they were the theatre. T. and R. will use means for obtaining these names and their interpretation; and I think they will confirm our conjecture that these roads were intended for a display of barbarous magnificence in hunting.

So the poet, engineer, and statistician go against current opinion. They favour an explanation based not the forces of nature, but on Gaelic names for the local topography. Of course they had no idea of the mismatch between geological and human timescales: in 1819 almost everyone believed that the Earth was only six or seven thousand years old, according to Christian teaching. The following half-century would bring enormous advances – especially by British geologists – in the identification and dating of geological strata by the fossils they contained, and in scientific estimates of the Earth's age. Today the Parallel Roads are known to mark the shorelines of an ice-age loch. Around 12,000 years ago a massive glacier formed a dam of ice behind which the loch formed, and as the glacier shifted it etched two more water levels on the hillsides of Glen Roy.

After enjoying a wonderful spectacle in glorious sunshine they returned to a late dinner. They were nearing the end of their stay in Fort William, and Southey airs some thoughts on the garrison town and its accommodation:

> The fort here is not so pleasantly situated as that at Fort Augustus, nor so picturesque in itself; but it is the only decent part of this place. The town is one long, mean, filthy, street; the inn abominably dirty; worse even, in this respect than the Field Marshall's, which was worse than any of our former quarters. We are in a spacious room, about 36 feet by 18, with an ornamented ceiling, from which the white-lime is peeling off. There are several panes of bulls-eye glass in the windows, two or three patched panes, and one broken one. The sashes are not hung – this indeed is the case everywhere. In one corner of the room is a bed, behind a folding screen. T. occupies this bed, and I have one upon the floor; we arranged matters thus the first night, rather than either of us would sleep in a double-bedded room, where the other bed was occupied by a gentleman and his son …

*A delightful scene in Fort William
(Geograph/Richard Dorrell).*

When the time came to leave they asked if their linen was ready and were told that the staff were *toasting* it. When Telford paid the bill, he gave a 20-shilling note to the poor girl who had acted as waiter and chambermaid (and probably also as chief cook). Southey was taken by 'the sudden expression of her countenance and her eyes' when she realised the huge tip was hers to keep. Their journey down the Great Glen had ended with an act of kindness.

To Glencoe, Inveraray, and Glasgow

See map on page 69

Travelling from Fort William towards Ballachulish, the party paused at the Corran Ferry, where Loch Eil meets Loch Linnhe. Rickman and Mitchell went across to inspect the pier on the opposite side, leaving the others to amuse themselves on the rocky shore. For Southey, a man of freshwater Lakeland lakes, it was quite an eye-opener. The tide was low and he was enchanted by a huge variety of saltwater creatures: barnacles, limpets, periwinkles, whelks, sea slugs, crabs, starfish, sea snakes, and jellyfish – 'I could not have enjoyed a more lively pleasure in all this if I had been five and thirty years younger, and I think Mrs R. and Miss Emma enjoyed it quite as much.' Amazingly, this is the poet's first mention of Miss Emma since they left Edinburgh five weeks earlier. Perhaps his lively pleasure was heightened by discovering that an attractive young lady shared his love of nature.

Arriving at North Ballachulish, they took a walk along the shore of Loch Leven, surrounded by superb mountain scenery:

> Yet even the mountains of Glencoe will not leave with me a more vivid recollection than a solitary sea bird, which while we were sitting on a little rocky knoll, dived into the water just below us, and when it emerged shook its wings, turned up its white breast, which actually seemed to flash like silver in the light, and sported so beautifully and so happily, that I think few sportsmen could have pulled a trigger to destroy so joyous a creature ...

> The evening was glorious. To the west the Linnhe Loch lay before us, bounded by the mountains of Morvern. Between those two huge mountains, which are of the finest outline, there is a dip somewhat resembling a pointed arch inverted; and just behind that dip the sun, which had not been visible during the day, sunk in serene beauty, without a cloud; first with a saffron, then with a rosey light, which embued the mountains, and was reflected on the still water up to the very shore.

Southey has changed gear. After a hectic week in the Great Glen spent among the engineering challenges of the Caledonian Canal, he is aware that Telford and Rickman have completed most of their professional inspections. The tour is entering its homecoming phase, and he relaxes – temporarily, at least – into poetic mode.

At Ballachulish, there were two small inns, one on each side of the ferry. Neither had enough rooms for the whole party, so Southey and Telford stayed on the near side while the others, accompanied by landau and gig, crossed over. The passage was notorious, and to many passengers frightening:

> It is a perilous ferry. The tide, having ten miles of lake [Loch Leven] to fill above the straight, presses thro' with great rapidity. The passage is impracticable with a strong westerly wind; with a strong easterly one very difficult. And there are

Loch Leven and the Pap of Glencoe from near Ballachulish (Geograph/M.J. Richardson).

no means for embarking horses or carriage with convenience, so that travellers who know the place always look onto this passage with apprehension.

The ferry remained a serious bottleneck on a major route from Glasgow into the Highlands until the 1970s, when it was superseded by a bridge. Previously the A82 went right round the shore of Loch Leven, a detour which could easily take a couple of hours on a bad day; modern travellers save time and nervous energy by crossing Loch Leven comfortably above the swirling tide race.

A long climb through Glencoe now beckoned horses, landau, and sightseers. Southey cannot help comparing its majestic scenery with his home country of Borrowdale and Buttermere; but he judges it wilder and barer, the mountains more serrated and cut into innumerable small ravines which wash stones down onto the road. It had become impossible to maintain a good road surface due to the 'rapid wreck of the mountains, which goes on, year after year, with increasing rapidity'. Luckily the contractor, warned that they were coming, had employed men the previous day to clear the way.

Soon after emerging from the majestic gloom of Glencoe they stopped at the King's House, a solitary inn some 1,200 feet above sea level:

The Three Sisters, Glencoe (Geograph/Sylvia Duckworth).

There were two beds in the room wherein we breakfasted. For the first time upon our journey, the house could supply no bread … there were, however, turkey as well as hen's eggs, a shoulder of lamb, and cream for the tea, which we had not found either at the Ferry or Fort William. Both here and at the Ferry there was handsome English china. Goats are kept here, and they make goat hams, which I was desirous of tasting … but they had not yet been smoked. The level on which the house stands may be considered as part of the Moor of Rannoch, the most elevated level in this part of the Highlands, from whence the waters flow in all directions, East, West, and North and South. It is a tract of peat and gravel with detached pieces and crags of granite. I saw a specimens of the lichen geographicus, which is so common and so beautiful in Cumberland.

The remote flatness of Rannoch Moor makes an extraordinary contrast with the peaks of Glencoe. I recall the magic of catching an overnight sleeper train in London and waking at the crack of dawn next morning amid the marvellous desolation of the moor, stretching away towards Buachaille Etive Mor at the head of Glencoe, followed, a few minutes later, by the diminutive Rannoch Station, second only in my affections to the one at Achnashellach, between Inverness and Kyle of Lochalsh.

Left: Rannoch Station (Geograph/Richard Webb).

Right: King's House Hotel near the head of Glencoe (Geograph/Richard Webb); and a train on the West Highland Line climbing up onto Rannoch Moor (Geograph/Alan Mitchell).

Southey often comments on political issues, and one of his favourites is the salt tax levied in Scotland. He has already mentioned an unintended consequence – in Banff, barrels of salted herring were exempted from the tax, encouraging the seller to substitute salt for fish: 'the more salt the worse for the fish, but the better for the seller. It seems that this exemption which is so well intended, and at first appears so just and unexceptionable, gives occasion to great frauds, smuggling, and evil in many ways'. He now discovers that all is not right in the West Highlands:

> A great contraband trade in salt is carried on upon these roads … A man who
> had formerly worked on the roads, but who found that the illicit salt-trade was a
> more gainful occupation, made a bargain at the Kings House to give six bolls of
> salt for a new cart, and deliver four of them at Tynedrum, and the other two at
> Ballachulish. The value of the cart was six or seven pound, the salt had cost him
> 7s. 6d. per boll, and he estimated the charge of delivering it at 10s. so the bargain
> on his side was a good one. On the other hand the purchaser sold it again for 2£
> per boll, and thus made more than an equal profit … The Excise officers give very
> little interruption to this trade, because the value of a seizure is far from being
> an adequate compensation for the trouble and risque of making it. There is great
> profit to be made by dealing in smuggled salt, and very little by seizing it. The
> Landlord of the Kings House took that Inn ten years ago, and had only a capital
> of 70£ to begin with. This year he has taken a large farm, and laid out 1500£ in
> stocking it.

Scotland's huge sea salt industry peaked in the early 19th century, a tempting product for government taxation and a notorious invitation to smugglers. Salt became known as 'Scotland's white gold', but the illegal trading was eventually wiped out by a repeal of the salt duty in the 1820s, after which cheap rock salt from the Continent flooded the market.

The landau continued 10 miles over Rannoch Moor to Inveroran, where Lord Breadalbane was building a new inn 'upon a poor mean scale, in a situation where a tolerable one would be very useful'; then another 10 miles to Tyndrum, once more in mountain country:

> The last three miles of the stage are on a descent, and unhappily they are – in
> Perthshire; therefore they formed a sad contrast to the fine smooth roads on
> which we had for three weeks been travelling: for the Perthshire roads are in a
> most barbarous and disgraceful state. We past [*sic*] a precipice down which a
> chaise had fallen some few years ago, with a gentleman, his wife, and a daughter
> in it; the gentleman received so much hurt that he died in two or three days, the
> wife had her thigh broken, the daughter alone escaped; the driver broke both
> legs, and one horse was killed.

They stayed overnight in Tyndrum, but for once Southey says nothing about the accommodation. He has lapsed into sombre mood, perhaps because the horses were exhausted after climbing up through Glen Coe to the King's House, and then to the top of Rannoch Moor;

perhaps because the moor had been cloaked in mist or drenched in Highland rain. In any case they had covered 35 miles through wild country which, to most travellers in 1819, would have seemed 'awful' in the original sense of the word. They were doubtless exhausted, and it comes as no surprise that Southey mentions a 'wretched assemblage of hovels' close to Tyndrum, and a country 'black and dreary, with high mountains on all sides; no cultivation except immediately about these hovels; no trees; the Fillan, a melancholy stream in the bottom'. He was probably recalling a delightful hour spent the day before in attractive female company on the shoreline of Loch Linnhe. The poor man could have done with a wee dram or, better still, a shared mutchkin.

His mood was hardly lighter next morning as they set out for Dalmally, 12 miles west of Tyndrum. Although no longer in Perthshire, he has another go at its roads, especially the ones running through Lord Breadalbane's estates, upon which his Lordship deserves to be driven 'from morning till night, till they should be completely reformed, at his charge'. Not that surface improvements would cure a fundamental fault in the old military roads now maintained by the Commission, which were 'needlessly steep and needlessly circuitous, thereby occasioning loss of time, and a grievous waste of exertion for the horses'. Generals Wade and Caulfeild had laid their roads out for troops on foot and horseback, not for a family landau. After 9 miles they descended into a valley which the road 'ought always to have followed'. The country now improved, the bright white tower of Dalmally church beckoned, and the mood lightened: 'The inn in this very pretty village is far better than any between it and Inverness.'

Immediately beyond Dalmally they came to the head of Loch Awe with its 'number of small islands, beautifully diversified with wood; the ruins of a Castle upon one', and turned south towards Inveraray. The countryside was now the gentlest since leaving Inverness, with 'fine trees in abundance of all the kinds which are common in Cumberland'.

After all Southey's disparaging remarks about Scottish landowners it is refreshing to find him enthusiastic about Inveraray, its castle, and the Dukes of Argyll:

Inverary [*sic*] is a small town, built by the last Duke, who spent a long life most meritoriously in improving his extensive estates and especially this fine place. The main street, terminated by a Kirk, reminded me of those little German towns, which in like manner have been created by small Potentates, in the plenitude of their power … At a little distance the appearance of these large buildings upon the shore, and the whole

The parish church in Dalmally, built in 1811 (Wikipedia/Trevor Littlewood).

surrounding scenery, bore no faint resemblance to a scene upon the Italian Lakes (Como more particularly) both in the character of the buildings, and the situation. The day was not favourable – a grey, Scotch, sunless sky; and the water of course grey also; but not lifeless, for there was just wind enough to keep up a sea-like murmur upon the stoney beach. The only bright part of the landscape was immediately about the Castle, where the grass under the trees was of a rich and vivid green. But even with the disadvantage of this sombre atmosphere, Inverary still, on the whole, exceeded anything which I have seen in G. Britain.

Praise indeed; and a welcome change from the previous day, when Glencoe and Rannoch Moor had so dampened his spirits.

Saturday, September 25. – The steam boat which has lately been started to ply between Glasgow and Fort William, and touch at the interjacent places, brings a great number of visitors to Inverary. As many as an hundred have sometimes landed there, to idle away more or less time, according to their means and leisure; many of them landing in the morning and returning in the evening. A revenue cutter is lying in the bay.

The Inveraray Castle that he saw had actually been built only a short time earlier: in 1744 the third Duke of Argyll, chief of Clan Campbell, had decided to demolish the existing building and start from scratch – not, this time, on traditional lines, but with a grandiose Gothic-style pile which would play host to many famous people, including Queen Victoria and her daughter Princess Louise, who had married a Campbell heir. Today, as in 1819, it is a much-visited tourist attraction.

Inveraray itself was also relatively new. In 1770 the fifth Duke set about rebuilding the town and engaged John Adam as architect. The final result, a delightful selection of 18th-

Left: Inveraray Castle (Geograph/Stuart Wilding).

Right: Loch Fyne from Inveraray. Clyde Puffer Vital Spark
awaits the next tide (Geograph/Steve Fareham).

century architecture, included houses for estate workers, a woollen mill and a pier to exploit the herring fishery on Loch Fyne.

'Herring' is a word guaranteed to get Robert Southey going – and by now he fancies himself an expert:

> The herrings of this Loch are generally reputed to be the best on the Western coast; tho' every Loch claims the superiority for its own; and all the Westerns insist that the herrings on the Eastern coast are so poor that they are fit for nothing but the West India market. But I who have revelled on herrings this season; so that this year of my life might be designated as the great Herring year – I who have eaten them with proper constancy, at breakfast, at dinner, and at supper also, when we supt, wherever they were to be had, from Dundee to Inverary – I as a true lover and faithful eater of this incomparable fish, am bound to deliver a decided opinion in favour of the herrings of Cullen above all others whatsoever.

How he could differentiate between the silver darlings of Cullen and, say, Banff is unclear; but he is absolutely clear that the open sea of the Moray Firth beats the lochs on the west coast; and in fairness we must admit that he has obtained plenty of practice.

Although awareness of sustainability issues has become commonplace in recent years, it was in extremely short supply two centuries ago. Southey seems ahead of his time:

> This fishery is an inexhaustible source of good and wholesome food, but not an inexhaustible source of commerce and wealth, as has been too often asserted by inconsiderate writers. The Dutch derived great riches from it, because they had the whole market to themselves, and were themselves a mere handfull of people. But what can be more unreasonable than to argue from this fact, that the same business can be pursued to any extent, and that as there is an infinite supply of herrings in the sea, so shall there be an infinite demand for them? Yet upon this assumption the Fishing Companies seem to proceed; and they will overstock the West Indies and the Levant with salted herrings, just as other speculators are overstocking Europe, Asia, and America with Manchester and Birmingham goods. The way to establishing a great, sure and lasting branch of business in this food is to convince our labouring people, and indeed all persons who have any regard for economy, that it is at once cheap, savoury, and wholesome. Sixpenny worth of advice and instruction concerning food and cookery, written in Franklin's manner, would do more good than all the Cheap Repository books and Evangelical pamphletts that have ever been dispersed.

We might argue with his first sentence, in that no fishery can be an inexhaustible source. Otherwise the awareness, especially coming from an early 19th-century poet, is laudable.

Leaving Inveraray they travelled round the head of Loch Fyne as far as the Cairndow Inn, then took the mountain road (now the A83) through the Arrochar Alps. Pausing at the summit, Southey noticed a seat and a stone 'bearing the beautiful inscription which all travellers have noticed – Rest and be Thankful'. They would have stayed to enjoy the mountain scene at leisure, but it began pouring with rain – everything was 'wild, great, simple, and severe'. But, unlike Glencoe, not 'terrible or savage'. The military road now descended towards the head of Loch Long, ending just short of the Arrochar Inn, 'and with it the power of the Commissioners. We entered Dunbartonshire, and the jolting was immediately such, that with one accord we pronounced the Dunbartonshire roads to be worse than the Perthshire'. Fortunately the inn was:

> more beautifully placed than any which I have seen either in Scotland or elsewhere – a large good house with trees about it, not a stone's throw from the shore, and with the high summit of the grotesque mountain abominably called the Cobler [*sic*], opposite and full in view.
>
> Sunday, September 26. – The comfortable accommodations at Arracher [*sic*], and the beauty of the grounds around it are now explained. It had been the seat of a Gentleman who outran his means, and was obliged to sell his estate, upon which the old Duke of Argyll, who wanted an Inn there, purchased it for that purpose. Coarse cotton sheets here – of Glasgow fashion sans doubt – I have learnt not to dislike them, upon this journey. Having only two stages for this day's work, we breakfasted before we set out.

They soon reached Tarbet beside Loch Lomond; and then, 'by a very beautiful, but in great part a very bad road', the inn at Luss, halfway down the western shore. Southey compares the outline of Ben Lomond to Skiddaw, but judges the rise steeper and the summit more rounded. They were soon entering prosperous country, 'once more among hedges and fields, villages and towns'. As the landau approached Dumbarton he espied 'a woman walking

Left: Rest and be Thankful, Arrochar Alps (Geograph/Richard Webb).

Right: Dumbarton Castle, undergoing a recent renovation on its 'remarkable rock' (Geograph/wfmillar).

barefoot, and with her bare legs exposed half way up, tho' she was expensively drest, and wore a silk spenser … glass-houses pouring out volumes of smoke … the remarkable rock on which the Castle stands'. Dumbarton appeared a neat town, and they put up at the Elephant and Castle, a large and comfortable inn whose main room was decorated with a portrait of an enormous ox that Southey recognised – it had been bred and fed by a friend at Halton Castle, Northumberland, who was also a donor of £3,000 to the Bible Society. With such anecdotes and chance remarks Southey continues to inform and amuse.

Mitchell was due back home in Inverness the following day. Southey had learned a great deal about his fellow traveller since they had met on the east coast, and now gives him a long and generous paragraph:

> Mr Telford found him a working mason, who could scarcely read or write. But his good sense, his good conduct, steadiness, and perseverance have been such, that he has been gradually raised to be Inspector of all these Highland Roads which we have visited, and all which are under the Commissioners' care, an office requiring a rare union of qualities – among others inflexible integrity, a fearless temper, and an indefatigable frame. Perhaps no man ever possessed these requisites in greater perfection than John Mitchell … No fear or favour in the course of fifteen years have ever made him swerve from the fair performance of his duty, tho' the Lairds with whom he has to deal have omitted no means to make him enter into their views, and do things, or leave them undone, as might suit their humour, or interest. They have attempted to cajole and to intimidate him, equally in vain. They have repeatedly preferred [*sic*] complaints against him in the hope of getting him removed from his office, and a more flexible person appointed in his stead; and they have not unfrequently threatened him with personal violence. Even his life has been menaced … In the execution of his office he travelled last year not less than 8800 miles, and every year he travels as much. Neither has this life, and the exposure to all winds and weathers, and the temptations, either of company or of solitude at the houses in which he puts up, led him into any irregularities or intemperance: neither has his elevation in the slightest degree inflated him. He is still the same temperate, industrious, modest, unassuming man, as when his good qualities first attracted Mr Telford's notice. Inverness is his home; he is a married man, and has several fine children.

Perhaps Southey exaggerates. He has clearly been affected by travelling in Mitchell's company in weather fair and foul, staying at inns equally variable, and has noticed the trust and admiration shown him by Telford. John Mitchell was undoubtedly an ox of a man, an extraordinary character with a devotion to duty that rivalled Telford's other great man-on-the-spot, Matthew Davidson. Sadly, both men died in harness: Davidson, at the age of 64, shortly before the landau and its occupants arrived in Inverness; Mitchell five years later, at the age of 45, still working his guts out for the Parliamentary Commission on Highland Roads, Bridges, and Harbours.

The travellers said farewell to Mitchell early next morning and took the Glasgow road alongside the River Clyde on which several new-fangled paddle steamers were plying their trade. Unfortunately, on their arrival in Glasgow the Buck's Head on Argyle Street failed to provide a welcoming fire for them, 'cold and hungry, at ten o'clock, on a wet morning' – a great disappointment, in stark contrast to the delightful inn at Arrochar and the comfortable Elephant and Castle at Dumbarton. But Southey has an explanation, and an anecdote:

> The Inns in large cities are generally detestable, and this does not appear to form an exception from the common rule. But it afforded what I cannot but notice as a curiosity in its kind unique, as far as my knowledge extends. In the "Commodité", which is certainly not more than six feet by four, there was a small stove, which as I learned from certain inscriptions in pencil on the wall, is regularly heated in the winter!

And why not? The Buck's Head may be reluctant to supply warmth in its largest rooms, but at least manages to do so in the smallest.

Our lover of mountain scenery is no admirer of large cities – even though he has previously tapped Edinburgh, Perth, and Inverness for medical advice about his 'volcano'. Glasgow does not even get this accolade, being 'a hateful place for a stranger, unless he is reconciled to it by the comforts of hospitality and society'. Better to reconnoitre the outskirts, and consult books for everything else. Actually they spent several hours in Glasgow, and he admits that Argyle Street is one of the best in Britain, and that the cathedral is 'the only edifice of its kind in Scotland which received no external injury at the Reformation'. Temporarily back in anecdotal mood, he derides church windows painted to imitate stained glass, and the tastelessness of seats 'so closely packed that any person who could remain there during the time of service in warm weather, must have an invincible nose. I doubt even whether any incense could overcome so strong and concentrated an odour of humanity.' And so he

The name survives: Buck's Head Buildings in Argyle Street, Glasgow (Geograph/Thomas Nugent).

continues, for another couple of pages, alternately critical and complimentary about a great city experiencing phenomenal growth as Scotland's industrial revolution gathers pace.

Glasgow marked the end of their six weeks in the Highlands, but not quite of their tour. Over the next few days the party continued south towards the English border, spending nights in Hamilton, Douglas Mill, Moffat (where they exchanged the trusty landau for two, faster, post-chaises), and Longtown near Gretna Green. On the way they called in at New Lanark, where Southey was fascinated to see the vast mills powered by the River Clyde, and to meet owner Robert Owen, who was conducting one of the industrial revolution's most famous experiments in paternalistic management of a huge resident workforce.

Telford said his own goodbye at Longtown and took the mail coach to Edinburgh. It was a poignant moment for Robert Southey:

> This parting company, after the thorough intimacy which a long journey produces between fellow travellers who like each other, is a melancholy thing. A man more heartily to be liked, more worthy to be esteemed and admired, I have never fallen in with; and therefore it is painful to think how little likely it is that I shall ever see much of him again – how certain that I shall never see so much. Yet I trust he will not forget his promise of one day making Keswick on his way to or from Scotland.

The poet reached home in time for dinner, finding 'all well, thus happily concluding a journey of more than six weeks, during which I have laid up a great store of pleasurable reflections'. He cannot possibly have imagined sharing them with admirers of his newfound friend, Thomas Telford, two centuries later.

For me, the greatest merit of Robert Southey's journal is its delightful account of an adventure among the Highlands and its people over a six-week period during which Telford was a tourist and sightseer as well as a professional engineer. Telford has often been painted as a man totally devoted to his work, but now we have seen him enjoying himself in wider landscapes and handling all sorts of socially demanding situations: equally at ease with Highland chieftains in their castles and rough innkeepers in unkempt premises; putting up with travel conditions that would have tried the patience and stamina of many a man half his age; spending two or three hundred hours in a landau packed with human bodies, male and female, and their luggage; sharing bedrooms with a man he has only just met, who has an open wound on his head that needs daily dressing; and finally ending up as the valued friend not of a fellow engineer but of a poet.

The far north

To Wick and Thurso

Mention Thomas Telford, and most members of the general public, if they know anything about him at all, will recall his famous suspension bridge over the Menai Straits in North Wales. A canal enthusiast may add the Pontcysyllte Aqueduct or the Caledonian Canal; a civil engineer, the road from London to Holyhead. But few are aware of his vast infrastructure programme in the Scottish Highlands, undertaken as engineer to the Parliamentary Commission on Roads and Bridges (subsequently including Fishing Harbours and Ports) set up in 1803–4. For its historical and social importance, and the huge amount of effort expended over a prolonged period, this programme is arguably the greatest achievement of his career.

Between 1806 and the early 1820s Telford's labour in the Highlands and Islands gave birth to over 1,000 miles of new roads, 280 miles of rebuilt military roads, 1,000 bridges, and more than 40 new and refurbished harbours and piers. Of course no man could have built these astonishing totals himself – but he took ultimate responsibility for them, making regular inspections of work in progress and keeping the parliamentary commissioners happy. As we have seen, he proved particularly adept at finding and appointing colleagues to act as his men on the ground, charged with overseeing the work to his designs and specifications, and reporting back on a regular basis. His 'Tartar', John Mitchell, the overall superintendent of works in the Highlands, was supported by a talented group of engineers appointed as local superintendents in six districts: Argyll, Badenoch, Lochaber, Ross-shire, Skye, and Caithness & Sutherland. The last of these, a vast and sparsely populated area of the 'Far North', was superintended by Thomas Spence.

It is hard to describe Telford's work in strict chronological order because, like other great 19th-century engineers, he was engaged on many diverse projects at any one time, dipping in and out as they progressed – often over many years. An alternative approach is to divide them into separate categories such as roads, bridges, and canals. A third is to arrange them as Robert Southey did in his journal – in geographical order, as he saw them. This has clear advantages for the modern tourist, and since I have become quite a fan of Southey's mix of engineering, history, geography, and personal anecdote, I have decided to follow his example. So we will now begin a tour of our own, visiting areas in the Highlands and Islands that Southey, the Rickmans, and Miss Emma Piggot failed to reach.

I propose starting at Fleet Mound, 8 miles north of Dornoch in the County of Sutherland, the point where Southey and co turned their landau around on 6 September 1819 and headed back south. The main A9 stretches before us towards Dunrobin Castle, then up the east coast to Wick, often in the presence of dramatic coastal scenery.

Two miles beyond Dunrobin we reach the small industrial town of Brora, with a notable war memorial in the form of a baronial clock tower. Like most towns along the east coast of Scotland, Brora was built around the mouth of a river flowing into the North Sea, and needed a good bridge. Old Brora Bridge, a single-span structure, had been built (and repaired) long before Telford's time. It is still in good condition, but reserved for pedestrians; traffic uses the 1930s road bridge a few metres upstream. In its time Brora has hosted coal mining, boat building, salt pans, fish curing, a lemonade factory, a wool mill, and a quarry that provided stone for London Bridge, Liverpool Cathedral and Dunrobin Castle. Thanks to its industry Brora was the first place in the north of Scotland to have electricity, earning it the nickname Electric City.

Brora is also well known to whisky enthusiasts for the single-malt produced at Clynelish on the northern outskirts of the town. The original distillery, like the settlement at Brora, was built by the Marquis of Stafford in response to James Loch's plans for a radical modernisation of the Sutherland Estates during the Highland clearances. Unsurprisingly, this included putting displaced crofters to work in the marquis's new business enterprises. By a strange coincidence the distillery was completed in 1819, the year of Robert Southey's tour, and became known as one of the Clearance Distilleries.

Left: Snowdrop time at Dunrobin Castle (Geograph/Andrew Tryon).

Right: Brora war memorial and, just beyond, the 1930s bridge over the River Brora (Geograph/Andrew Tryon).

Left: Old buildings of the Brora Distillery at Clynelish (Geograph/Andrew Wood).

Right: The Emigrants' Statue in Helmsdale (photo: Brian Caddick).

Eleven miles beyond Brora we come to Helmsdale, which grew as a planned village after 1814, the notorious Year of the Burnings. It was laid out on a grid pattern by Sutherland Estates in the hope of creating a community that could survive by its members joining the rapidly growing herring industry. However many former tenants of the inland straths found themselves sailing away to Canada and America instead. Their plight is commemorated by a fine modern statue overlooking the village.

As Robert Southey makes clear in his journal, Telford and his chief assistant, Mitchell, were heavily involved in bridge building and road making north of Inverness in the early 1800s. The modern A9 follows much of the route they pioneered. One important contribution was a bridge at Helmsdale, completed in 1811 and still in use. An unusual Telford design of two segmental arches, each of a 70-foot span, with triangular cutwaters and a level road deck, it crosses the River Helmsdale at the inland end of the village street.

Helmsdale, like Brora, was developed around the mouth of a river; but unlike its neighbour, which only had a small harbour, Helmsdale was given a generous one. Started in 1818 and extended in 1823 – no doubt with advice from Telford – it became home to a large herring fleet. It is still important for the life of Helmsdale, with leisure craft and fishing vessels that give the feel of a working harbour. And just to show that today's Helmsdale, after a sad 19th century, is a far happier and healthier place, the village proudly puts on its own Highland Games in August each year.

The first time my wife and I came this way, we diverted from the A9 to take the A897 inland through the Strath of Kildonan, which had been ruthlessly cleared of inhabitants in both 1813 and 1819. We drove alongside the River Helmsdale and the single-track railway that leads, by a very roundabout route, to Wick and Thurso; and passed two small tributaries of the river that became famous in 1869 – not for trout, nor even salmon, but for *gold*. The precious metal was discovered on the Suisgill Estate by one Robert Nelson Gilchrist, a native

Telford's bridge over the River Helmsdale (Geograph/James Allan).

Helmsdale Harbour (Geograph/James Allan).

of Kildonan who had spent 17 years in the goldfields of Australia and brought his expertise back home. Hundreds of prospectors flocked to set up camp and try their luck in the Great Sutherland Gold Rush. Precious-metal fever proved short-lived, but recreational gold panning is still permitted and visitors are welcome to try their hand. I gather that large nuggets are so rare that you are unlikely to pay off the mortgage; but sploshing about in the burn is said to be great fun for children between the ages of 8 and 80.

Our own excursion was not a search for gold, but for a new RSPB reserve then being established at Forsinard, about 25 miles northwest of Helmsdale. We had read of a scandal developing around a government tax break that attracted rich individuals (including certain celebrities who should have known better) to plant large tracts of forest in the Flow Country of Sutherland, threatening its unique ecology. Now, the RSPB was saving the ecosystem by encouraging supporters to purchase small blocks of as yet unspoiled blanket bog at what was, by most standards, an incredibly low price, and we had in mind a donation in memory of a family member who had been a keen birdwatcher. Our visit was amply rewarded: the Flow Country turned out to be wonderful, the reserve (now known as Forsinard Flows) inspiring. According to the current RSPB website:

> Forsinard Flows is part of a vast expanse of blanket bog, sheltered straths and mountains known as the Flow Country. The Flow Country is one of Scotland's most important natural treasures and the RSPB looks after more than 21,000 hectares of it. The RSPB has been working to protect the landscape here for more than 20 years.

Forsinard itself consisted of little more than a railway station and a hotel where we had pre-booked a three-night stay. It turned out be be a bit run down and so we wondered whether

In the Flow Country (Paul A. Lynn).

two nights would be enough, but soon discovered that it had the fishing rights on several local lochs, with rods, reels, and flies available for the asking. One crazy visitor had just arrived from Surrey after a semi-continuous 12-hour drive, solely for the fishing, and encouraged us to have a go. So next morning we set out at dawn, had one of the lochs entirely to ourselves and, in spite of minimal casting skills, landed a couple of brown trout. What really changed our minds about the Forsinard Hotel was the friendliness of its staff, an unsolicited offer from the chef to fry our trout in butter for breakfast, and the company of guests with a shared love of wild places in unfashionable locations.

Returning to Helmsdale we rejoined the modern A9 northbound, and soon began a dramatic climb up and over the Ord of Caithness, a granite headland on the boundary between Sutherland and Caithness that rises 650 feet above the North Sea. Historically the track over it was the main land route into Caithness, and 'traversed the crest of its stupendous seaward precipices at a height and in a manner most appalling to both man and beast'. Even Telford's improved road, completed in 1811, had sharp bends and stiff gradients that were feared by riders, and coachmen and their passengers.

Progressing north, Telford continued bridge building over burns that spilled down into the North Sea – preferred sites for coastal villages. Examples are Berriedale (where he built a pair of single-span rubble bridges near the confluence of Langwell Water and Berriedale Water); Dunbeath; and Latheronwheel.

Berriedale is also notable for one of Telford's Highland churches, designed and built under his supervision in the late 1820s. Funded wholly or partly by the government to encourage churchgoing by restless Highlanders displaced from their traditional settlements, many of the buildings are still visible today. Some are active parish churches; others have been converted to secular uses (Berriedale's is now a charity); and a few have been abandoned to the elements. We shall meet a number of examples on our travels, and a discussion of them at the end of the book.

Dunbeath Castle, a well-known landmark (especially to sailors) some 17 miles beyond Helmsdale, has 15th-century origins and gazes seaward from a dramatic clifftop. The village of Dunbeath has a very attractive small harbour and is doubly important to me: first, for a historic mill building, the HQ of an independent Scottish publisher who has accepted several of my manuscripts; and second, for its well-preserved Telford bridge, completed in 1813.

Two miles further on we come to Latheronwheel, another village with a Telford bridge – but this time, it seems, semi-derelict and abandoned. In any case I cannot find a clear image of it, so am including an alternative to underline the importance of packhorse bridges before the days of Telford and Mitchell. As a local commented, 'Upon this rapid burn Patrick Dunbar of Bowarmadde has erected this year [sometime between 1717 & 1732] a stone bridge of a large arch, which will be of great use not only to the whole parish but to all that travel that road.' You can find it down near Latheronwheel's harbour.

Latheron, just beyond Latheronwheel, is where the modern A9 leaves the coast and strikes north to Thurso. But we will continue along the coast road (now the A99) towards Wick, soon coming to Lybster, the final village in a close-spaced set of four. An extremely long, straight main street harks back to the Victorian era and Lybster's extraordinary growth during the great herring boom. At its peak the village boasted Scotland's third-largest fishing

*Telford's 1813 bridge over Dunbeath Water, dwarfed by
the modern A9 bypass (Geograph/John MacKenzie).*

Dunbeath Harbour (Geograph/John MacKenzie).

Left: The 18th-century packhorse bridge at Latheronwheel (Geograph/Jim Matthews).

Below: Lybster's harbour (Geograph/ Ian Taylor), and restored buildings (Geograph/valenta).

fleet, with an accompanying population of fishers, gutters, coopers, and net menders. Today, the impressive harbour presents a delightfully peaceful scene – unlike the frenzied activity that signalled the arrival of a major catch of 'silver darlings' in its heyday. Restored buildings house one of the area's major attractions – an exhibition portraying Lybster's remarkable history, geology, flora, and fauna, with hands-on activities and refreshments. One end of the building has been set up as a Victorian smokehouse.

And now for something very different, and much larger. Wick, 13 miles beyond Lybster, has a current population of about 7,000 and a long history as one of Europe's premier herring ports. Telford was heavily involved in its expansion, and to put this in context I should like to recap a little.

The British Fisheries Society, set up in 1786, designated Tobermory on the Isle of Mull and Ullapool in Ross and Cromarty as fishing ports, and Telford gave advice on the design of their harbours. But when the normally plentiful herrings decided to go elsewhere the society turned its attention to Wick, a village on the north bank of the River Wick in Caithness. A major priority was a decent bridge – prior to 1800, travellers from the south could only access the village by a footbridge of 11 pillars connected by wooden planks.

Not surprisingly Telford came up with a design for a three-arched masonry bridge. But his contribution to Wick went much further, mainly because his patron and friend from the Shrewsbury days, Sir William Pulteney, had a finger in a great many pies – including the British Fisheries Society. Pulteney had admired Telford's work as an architect in Shropshire,

The present Bridge of Wick, a wider version of Telford's three-arch bridge, was completed in 1877 (Geograph/John Lucas).

including his church in Bridgnorth and Laura's summer house in the grounds of Shrewsbury Castle, and duly appointed him architect of a new town and harbour for the society on the south bank of the River Wick. Telford's design included everything needed by a self-contained fishing community. Following Sir William's death in 1805, it was named Pulteneytown.

Pulteneytown is considered by many to be Wick's hidden gem, a fine example of Telford's talent as an architect and planner. Today, Wick and Pulteneytown, on opposite sides of the river, make up what is generally known simply as Wick. The former has a busy shopping centre and 'the World's shortest street', 6-foot-long Ebenezer Place at the corner of River Street and Union Street. Pulteneytown can still be imagined as the heart of a 19th-century fishing port. And if you have time to spare, consider a visit to the Wick Heritage Museum, perhaps followed by a wee dram of Old Pulteney from the famous distillery, founded in 1826.

Telford's final piece of roadmaking in the far northeast of Scotland was between Wick and Thurso, the route of the modern A882/A9. On the way he built a handsome bridge over the Wick River at Watten. Approaching the village of Watten today it is fairly easy to miss Achingale Bridge – but not if we keep a lookout for an attractive welcoming stone at the roadside, decorated with a quirky combination of clock face, pendulum, three trout, and some sort of electrical contrivance. A little further on, the parapet of Telford's bridge, completed in 1817, is just visible. Actually it was widened on the north side in 1933, but its character survives, with a 90-foot span supported by three segmental arches with triangular cutwaters.

Wick Harbour: from the air (Geograph/Stanley Howe); and at sea level (Geograph/Dorcas Sinclair).

It is the last substantial Telford bridge on our present route, and shortly after crossing it we reach the centre of Watten.

Most British villages have interesting histories and personalities, and Watten is no exception: it was put on the 19th-century map by Telford's infrastructure, and one of the houses at the village crossroads started life as a tollhouse on his new road; nearby Loch Watten is the largest body of water in Caithness, famous for its trout fishing; during World War II the village hosted 'Britain's most secretive POW camp' and some very prominent Nazis; and

A pot still at the Pulteney Distillery (Wikipedia/ Lakeworther).

Watten's welcoming stone. The parapets of Telford's Achingale Bridge are visible just beyond the 30 mph sign (Geograph/ Paul Simonite).

The former tollhouse in Watten (Geograph/Paul Simonite).

last but not least, Watten was the birthplace of Alexander Bain, the 19th-century inventor of a pendulum-regulated electric clock. You can find a stone monument to him outside the village hall. Now we know why the village's welcoming stone is the way it is, and why the hotel opposite the tollhouse is called The Brown Trout.

The arrival of Telford's road in Thurso hastened the town's expansion – but not to the extent of a large harbour and fishing fleet, because it overlooks the Pentland Firth. In the age of sail the firth had a dreadful reputation among sailors and fishermen, especially when its swirling tidal

Left: Thurso promenade and beach (Geograph/Christopher Hilton).

Right: Scrabster Harbour (Geograph/Andrew Curtis).

races were met by a contrary gale. Thurso had to wait until 1841 for a deep-water harbour, by which time steamships, far less vulnerable to wind and tide, were in regular use. The harbour was built just outside the town at Scrabster, and has recently been developed into a flourishing gateway to Orkney, Shetland, and Scandinavia. Thurso, the most northerly town on the British mainland, has about 7,000 inhabitants; on a fine day with light winds its promenade is a delight, with sweeping views across the Pentland Firth towards the Orkney Islands.

Bonar Bridge to Tongue

Telford had one more major route to complete in the far north, the 45 miles due north from Bonar Bridge to Tongue, via Lairg and Altnaharra. The coastal village of Tongue lies about halfway between Thurso and Cape Wrath, and we will now follow his route – which is essentially that of today's A836.

Starting from Bonar Bridge, we travel 11 miles north to Lairg, beyond which the road (much of it single-track) passes through one of the remotest regions of Scotland. The scenery has its own stark beauty and there is little traffic even in summer. The stretch between Lairg and Altnaharra is heavily forested for the first 10 miles, but opens up beyond a summit at the Crask Inn to reveal fine views towards Ben Klibreck. A steady descent towards tiny Altnaharra (the site of a Met Office weather station that regularly records Scotland's lowest winter temperatures) brings Loch Naver into view. Just beyond the straggling village, we cross the River Mudale on an impressive Telford bridge, built four years before the road was completed in 1819.

About 4 miles beyond Altnaharra the northern landscape expands wonderfully, with views towards Ben Loyal, 'the Queen of Scottish Mountains'. We soon meet another Telford bridge, set in a remote valley at Inchkinloch, followed by a delightful run alongside Loch Loyal. Halfway along the loch at Lettermore a miniature bridge, presumably also by Telford, has been bypassed and abandoned to grass.

Telford's bridge at Altnaharra (Geograph/John Ferguson).

Inchkinloch bridge, still in use; and Lettermore, now abandoned (Geograph/Richard Webb).

Loch Loyal (Geograph/Richard Webb).

Many years ago my wife and I had a walking holiday in the far north, with Ben Loyal in our sights. We pitched our tent by a small burn and a miniature road bridge, presumably another Telford. No living thing, human or animal, appeared that afternoon or evening. We retired to our tent feeling secure in a vast landscape – until about four in the morning when heavy breathing was accompanied by tugging on guy ropes. A herd of Highland cattle with their exceedingly large horns had joined us, and although we waited an hour or so for them to clear off, they were obviously as contented with their pitch as we had been with ours. Increasingly anxious that they would demolish our accommodation with us inside, we debated whether to stay or flee semi-clothed to our car about 100 yards away. The strong metal box won the vote. The beasts eventually departed, leaving two dishevelled humans to salvage their belongings and their pride, and think hard before wild camping again in a Highland wilderness.

Wild Camping in Sutherland (Paul A. Lynn).

Ben Loyal from the Kyle of Tongue (Paul A. Lynn).

Kyle of Tongue. The long low bridge connecting Tongue village to the west side of the estuary was completed in 1971 (Geograph/Graeme Smith).

Ben Loyal is best seen from the Kyle of Tongue beyond Tongue village. At 2,506 feet (764 metres), it ranks not as a Munro (over 3,000 feet) but as a Corbett (named after the first person to climb all Scottish peaks above 2,000 feet). But, as we discovered, its isolation and dramatic rise above the sea level in the kyle give it a presence far above many a Munro, with wonderful views from its several summits over moorland and sea.

The beauty of this area belies a troubled, and occasionally violent, history. The Kyle of Tongue became famous for a battle between a Jacobite treasure ship and two ships of the Royal Navy in 1746. The ship's crew tried to slip ashore with their gold, but were caught by navy personnel and local people loyal to the British Crown (contrary to popular belief, Bonnie Prince Charlie was not universally loved and supported in the Scottish Highlands). By the early 1800s many Highlanders evicted from inland areas had been housed in Tongue and nearby coastal settlements. Bettyhill, about 12 miles along the tortuous coastal road from Tongue, was set up by Elizabeth, Countess of Sutherland (and named after herself) to house tenants evicted from Strathnaver in 1814, the Year of the Burnings. It is salutary to remember that Telford was busy building roads and bridges in the far north during this period, and that the handsome bridge we have just crossed at Altnaharra was completed in 1815. We may hope that the hundreds of men who laboured to bring the road up from Bonar Bridge felt proud of an achievement that would enhance the lives of the Highlanders who remained. Robert Southey wrote a thoughtful commentary on all this, balancing the urgent need for new Highland infrastructure against the cruel disruption to families displaced by sheep; and Telford, like his friend, was undoubtedly aware of the issues. Perhaps we should spare a thought for the area's history as we explore and enjoy its wonderful scenery and pristine environment.

The village of Tongue (Wikipedia/Florian Fuchs).

Roads to the Isles

Thomas Telford was an unstoppable multitasker. From 1804 onwards the Caledonian Canal demanded a huge amount of his attention, yet he was also devoting time and energy to the programme of Highland roads sanctioned by the parliamentary commissioners – and none more urgent than several that were to run from the Great Glen towards the west coast and the Hebridean islands. They would pass through regions steeped in Gaelic language, music, and tradition, former hotbeds of Jacobite sympathy where historic links with Catholic France and Spain rivalled those with Protestant England. They were mostly untouched by General Wade's military roads. Unemployment, exacerbated by the clearances, was causing poverty and emigration. Not surprisingly, Parliament wanted the country opened up and developed.

But before we travel the routes of these roads ourselves, I should say a little more about the state of the Scottish, and especially the West Highland, roads before Telford started improving the existing ones and driving new ones through virgin territory. And for this I turn once again to one of my favourite Victorian authors, Samuel Smiles, and his *Lives of the Engineers*.

According to Smiles, the agricultural progress that was well under way in the Lowlands by the 1790s had barely touched the West Highlands – 50 years after Bonnie Prince Charlie's rebellion. And the physical separation of the inhabitants, which was primarily due to the lack of roads and bridges, had profound social consequences:

> The native population were by necessity peaceful. Old feuds were restrained by the strong arm of the law, if indeed the spirit of the clans had not been completely broken by the severe repressive measures which followed the rebellion of Forty-five. But the people had not yet learnt to bend their backs, like the Sassenach, to the stubborn soil, and they sat gloomily by their turf-fires at home, or wandered away to settle in other lands beyond the seas. It even began to be feared that the country would soon be entirely depopulated … The poverty of the inhabitants

rendered the attempt to construct roads – even had they desired them – beyond their scanty means.

In 1802 Telford was asked by the government to survey the Highland infrastructure, and his report, submitted to Parliament the following year, declared that the existing military roads 'were altogether inadequate to the requirements of the population, and that the use of them was in many places very much circumscribed by the want of bridges over some of the principal rivers'. Among many recommendations, new roads were needed in the West Highlands to allow 'ready communication from the Clyde to the fishing lochs in the neighbourhood of the Isle of Skye'.

The new roads, bridges, and other improvements suggested by the engineer, excited much interest in the north. The Highland Society voted him their thanks by acclamation; the counties of Inverness and Ross followed; and he had letters of thanks and congratulation from many of the Highland chiefs. "If they will persevere, says he [Telford], with anything like their present zeal, they will have the satisfaction of greatly improving a country that has been too long neglected. Things are greatly changed now in the Highlands. Even were the chiefs to quarrel, devil a Highlandman would stir for them. The lairds have transferred their affections from their people to flocks of sheep, and the people have lost their veneration for the lairds. It seems to be the natural progress of society; but it is not an altogether satisfactory change. There were some fine features in the former patriarchal state of society; but now clanship is gone, and the chiefs and people are hastening into the opposite extreme. This seems to me to be quite wrong."

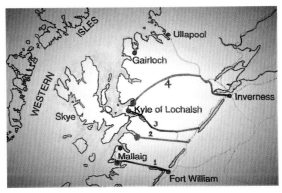

Telford's roads from the Great Glen to the fishing lochs opposite the Isle of Skye: (1) Fort William to Arisaig (later extended to Mallaig); (2) Invergarry (near Loch Oich) to Kinloch Hourn; (3) Invermoriston (near Loch Ness) to Kyle of Lochalsh; (4) Inverness to Strome Ferry.

Assuming Smiles is reporting Telford's words accurately, these comments about the lairds and their tenants are extremely revealing and presage the large-scale Highland clearances that were to follow. As he returns to his native land, our engineer is well aware of the state of Highland society, and is concerned about it. As a man in tune with the Scottish Enlightenment, he believes passionately that improvements in Highland infrastructure are sorely needed, not least to benefit the fishing industry, and will do all he can to help; but there are troubling social issues for which the lairds are as blameworthy as their tenants.

Telford drove four roads westwards from the Great Glen to link central and southern Scotland with the fishing lochs opposite Skye, the largest island of the Inner Hebrides. Their approximate routes are shown on the accompanying map. At various times the romantic sobriquet 'Road to the Isles' has been applied, with varying degrees of conviction, to all of them. It is now a generic term, used with relish in books, newspapers, and tourist guides.

1 Fort William to Arisaig

'Road to the Isles' is most commonly applied to the route from Fort William to Arisaig and Mallaig, from where ferries sail for Skye and several smaller Hebridean islands. Back in the 18th century the route had originated in Perthshire, then crossed Rannoch Moor and headed for Fort William. In the early 19th century Telford drove it on to Arisaig on the west coast, through country intimately associated with the adventures of Bonnie Prince Charlie. His route was very similar to that of the modern A830, one of the most scenic in Scotland. The West Highland Line, beloved by connoisseurs of great railway journeys, also runs this way.

Many people think of the Highlands as mountain, moorland, and heather, but Telford's road from Fort William started along the irregular shoreline of Loch Eil, intermittently covered by birch, rowan, and ancient oak – not the easiest terrain for a 19th-century roadmaker. Much of his route was up, down, and curvy, and remained single-track long after horse-drawn carriages had been overtaken by motor vehicles. In the 1960s the West Highland Line was initially marked for closure by the much-maligned Dr Beeching, but was granted a stay of execution because the A830 was judged too slow and dangerous for a substitute bus service. Then in 1991 local people blockaded the road to highlight its delapidated condition, and as recently as 2007 one of its sections achieved the dubious distinction of a 1 star, i.e. dangerous, rating from the Institute of Advanced Motorists, but these ratings have now been discontinued. Final upgrading in 2009 eliminated the last single-track section between Lochailort and Arisaig, but many older drivers will never forget the blind bends and summits of Telford's parliamentary route.

Bridge, probably Telford's, on the Road to the Isles at Fassfern, near Loch Eil (Geograph/Gordon Brown).

It is time to start our own journey along the A830 from Fort William. About 6 miles beyond Banavie the road and railway begin a straight and level run alongside Loch Eil, making clear that we are bypassing Telford's twists and turns and can expect few reminders of his original route. But halfway along the loch a narrow right turn towards Fassfern offers a short and rewarding diversion: we soon come to a lively burn spilling from the hills, crossed by an ancient stone bridge described by Historic Environment Scotland as 'probably Thomas Telford, 1803–4, single span, slightly humped; a back rubble bridge, dressed rubble arch ring, and low parapet, with widely splayed approach at west'. Homage paid, we can return to the modern A830.

A further reminder of Telford's original route may be found about 2 miles beyond the head of Loch Eil, as we continue along the floor of a broad valley. On the right, a short abandoned section of the old road leads to a derelict bridge across the Dubh Lighe (Dark River).

The fast new road passes under the railway line and we soon find ourselves in Glenfinnan, one of the most emotive places in the Scottish Highlands. It has a great deal to offer: a famous monument to Bonnie Prince Charlie at the head of beautiful Loch Shiel; an impressive railway viaduct; and a delightful station on the West Highland Line, complete with its own museum and tastefully restored dining car.

On 2 August 1745 a French privateer landed Bonnie Prince Charlie and seven supporters on Eriskay, a small island in the Western Isles. They reached the Scottish mainland in a rowing boat and came ashore at Loch nan Uamh about 14 miles west of Glenfinnan. The prince raised his standard at the head of Loch Shiel on 19 August and boldly claimed the Scottish and English thrones in the name of his father James, 'the Old Pretender' (whose claim to the throne of Scotland rested on his being the eldest son of Scotland's James VII / England's James II, deposed in 1688 for his Catholicism, in favour of his Protestant daughter, Mary). Eight months later the Jacobite rebellion ended in brutal failure at the Battle of Culloden near Inverness, and the prince fled to evade government troops. He was famously sheltered by Flora MacDonald on the Isle of Skye before escaping aboard a French frigate at the same Loch nan Uamh, never to set foot on Scottish soil again. Hundreds of Jacobite enthusiasts gather

The Glenfinnan Monument to Bonnie Prince Charlie at the head of Loch Shiel (Geograph/ Malc McDonald).

at the Glen Shiel monument on 19 August each year, to remember and lament his ill-fated adventure.

Our next attraction, Glenfinnan Viaduct on the West Highland Line, is a mile or so inland from the head of Loch Shiel. Designed by Robert McAlpine, it was constructed in 1897–8 and opened to passenger traffic in 1901. The head of the firm earned the nickname Concrete Bob for his pioneering use of mass concrete in the 21-span, 416-yard, curvaceous structure – the longest concrete railway bridge in Scotland. Soaring 100 feet above the River Finnan, it was built on site by pouring concrete into formwork, without metal reinforcement. Telford, who had pioneered the use of cast iron in his Pontcysyllte Aqueduct a century earlier, would surely have been fascinated.

A plaque beneath one of the viaduct's arches reads:

GLENFINNAN VIADUCT CENTENARY 1897–1997

THIS OUTSTANDING VIADUCT, A PIONEER WORK IN MASS CONCRETE,

WAS CONSTRUCTED

JULY 1897– OCTOBER 1898 AT A COST OF £18,904

CONTRACTOR – ROBERT McALPINE & SONS, GLASGOW

ENGINEERS – SIMPSON & WILSON, GLASGOW

PLAQUE UNVEILED 12 JULY 1997

BY

THE HON WILLIAM McALPINE, Bt.

The viaduct is still very much alive, supporting regular railway services between Fort William, Arisaig, and Mallaig, including excursions by The Jacobite steam train filled with enthusiastic passengers – many of them young fans of Harry Potter and the Hogwarts Express.

Left: The Jacobite crosses Glenfinnan Viaduct (Geograph/Ian Taylor).

Right: The Jacobite draws into Glenfinnan Station (Geograph/Malc McDonald).

Glenfinnan and its station have special memories for me – and for rather different reasons. I first came this way many years ago during a walking holiday with an old school friend, intending a long hike over the mountains at the head of Glen Finnan, down into Glen Pean and along the shoreline of Loch Morar to the coast at Mallaig. But the clouds came down, it started to pour – and failed to stop. We decided not to risk a trackless mountain route among Munros, so retreated to Glenfinnan Station and took the train. Soaked and disconsolate, we sheltered in a café on the quay at Mallaig where, quite by chance, we met two German girls who were hitch-hiking to Skye. I remain forever indebted to that four-day Highland downpour and the chat over a cup of coffee that changed my life for ever – and immeasurably for the better.

Beyond Glenfinnan the A830 enters the region known as Moidart and Arisaig, often referred to as the Rough Bounds. Historically isolated, much of it was only accessible by ferry or on foot until the 1960s. After 5 miles among mountains we pass along the north shore of freshwater Loch Eilt, leaving the south shore to the railway, and soon reach the head of saltwater Loch Ailort, our first glimpse of the western sea. That glimpse becomes a vista 2 miles further on at Loch nan Uamh, the site of Prince Charlie's landing and subsequent embarkation, commemorated by a Prince's Cairn erected near the shoreline in 1956.

The impressive railway viaduct at Loch nan Uamh has its own claim to fame. For many years a legend persisted that a horse and cart had fallen into one of Glenfinnan Viaduct's hollow piers during construction. A fisheye camera was inserted in 1987, but nothing was found. The local rumours were so persistent, however, that the search was transferred to Loch nan Uamh, and was finally rewarded in 2001.

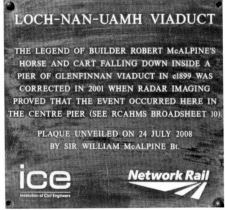

The railway viaduct at the head of Loch nan Uamh, scene of a bizarre discovery (Geograph/ M.J. Richardson).

The Prince's Cairn at Loch nan Uamh (Geograph/Gordon Brown).

Arisaig House (Geograph/Jim Bain).

The remoteness of the Rough Bounds proved ideal for covert military exercises during World War II, when the Special Operations Executive (SOE) established a Scottish base at Arisaig House, a mile or so beyond Loch nan Uamh at Druimindarrach. Specialising in Commando skills, the base provided training for around 3,000 agents destined to serve mainly in occupied Europe. The single access road to Arisaig made it easy to seal off, away from unwanted attention: the railway was convenient for staff and legitimate visitors. Security was rigorous, and even locals had to carry passes for movement in and out of the Special Protected Area. All this is hard to imagine as we approach Arisaig along a scenic coastline of white beaches with wonderful views to the islands of Eigg, Rum, and Skye.

Rum seen from Eigg (Paul A. Lynn).

White sands of Morar, with Eigg and Rum in the background (Geograph/Anthony O'Neil).

*Mallaig Harbour in its heyday
(Geograph/The Carlisle Kid).*

Telford's road stopped at Arisaig but was subsequently extended to the villages of Morar and Mallaig, the latter founded in the 1840s by Lord Lovat, who 'encouraged' his tenants to take up herring fishing as a way of life. The extension formed part of the A830 until recent times, but is now labelled the B8008 – a surprising addition to the classified routes of Scotland. As a result, you can now enjoy a breathtaking minor coastal road without having to worry about fast-moving traffic.

The population and local economy expanded rapidly in the 20th century with the arrival of the railway, which helped the fishermen of Mallaig sell their catches fresh in Glasgow, and even London. By the 1960s Mallaig was one of the busiest herring ports in Europe, and prided itself on its kippers. Today only one traditional smokehouse remains, but I am told the products are as good as ever.

2 Invergarry to Kinloch Hourn

See map on page 169

Telford's second road from the Great Glen to the west coast was roughly parallel with the first but about 15 miles further north. Starting at Invergarry near Loch Oich (the summit level of the Caledonian Canal), it ran about 35 miles past Loch Garry and Loch Quoich to Loch Hourn, a dramatic sea loch opposite Skye. The modern A87 follows a similar route for the first 7 miles, then climbs away northwards from Glen Garry towards Bun Loyne and the Cluanie Inn. Telford's original route, now classified as the C1144 and known as the Kinloch Hourn Road, carries on towards the west coast. Narrow and mostly single-track with passing places, it is reputedly the longest no-through road in Britain – a fascinating excursion for Telford enthusiasts. But before starting our own journey, it is worth recapping the story so far.

You may remember that Telford first travelled this way in 1801, to check that Lochs Quoich, Garry, and Oich could provide the Caledonian Canal with a plentiful water supply, and to explore the country around and beyond Loch Hourn. The road's starting point at Invergarry figures prominently in Robert Southey's journal of 1819: when Southey, Telford,

A wet morning in Glen Garry (Paul A. Lynn).

and Rickman called in on the local chieftain, Alexander MacDonell of Glengarry, they were received 'with much civility and satisfaction', but were left in no doubt that he was a haughty and unrepentant Jacobite, a force to be reckoned with. Eighteen years earlier he had proved highly awkward to Telford and Rickman over the land needed for the canal, and presumably drove a hard bargain over the new road, which passed through his extensive estates – land whose inhabitants over the years suffered more than their share of Highland clearances. We might bear all this in mind as we start our journey through Glen Garry, where former Highland dwellings are occasionally marked by melancholy piles of stones.

Telford's Kinloch Hourn Road leaves the A87 about halfway along the shoreline of Loch Garry, and continues westwards towards Tomdoun and Kinloch Hourn. Actually the A87 also continued as far as Tomdoun until the late 1950s, then turned sharply north towards the Cluanie Inn. But the damming of Loch Loyne by the North of Scotland Hydro-Electric Board in the 1950s caused flooding, the road was abandoned, and the A87 was re-aligned along its present route. All this gives me a welcome excuse for a diversion.

Many years ago we stayed at the delightful hotel in Tomdoun, where my wife learned that chanterelles – one of her favourite mushrooms – were abundant 'along an abandoned stretch of road nearby'. Somewhat gingerly, we ventured along it, with absolutely no idea of its Telford provenance or subsequent history. We were eventually stopped by floodwater

Tomdoun Hotel
(Geograph/David Purchase).

and struggled to turn the car round on the narrow, crumbling roadway; but at least we returned with a fine basketful of chanterelles – a star turn on the hotel menu that evening.

A fascinating set of photographs taken in 2008 shows the state of the abandoned Tomdoun-Cluanie road and its bridges which, as we had discovered, are usually submerged below the floodline. In September of that year Loch Loyne was reduced to exceptionally low levels, giving walkers a rare opportunity to examine miles of crumbling Telford infrastructure. As an engineer interested in renewable energy for more than 30 years, I am generally a fan of hydro-electricity, including the large scheme that includes Loch Loyne and several others in the area.

The Great Glen Scheme (sometimes called the Garry-Moriston Scheme) was completed in 1962 with a total generating capacity of 113 MW. Situated in one of the wettest parts of the UK, it has an annual rainfall of about 120 inches and a series of capacious lochs to collect it. The southern part of the scheme includes Loch Quoich, Loch Garry, and Loch Oich – the chain of lochs that Telford decided could guarantee a generous water supply for the Caledonian Canal. Today Loch Quoich acts as the scheme's main storage reservoir, retained by the largest rockfill dam in Scotland, 1050 feet (320 metres) long and 125 feet (38 metres) high. Water is sent by tunnel to the 22 MW Quoich power station. Downstream, Loch Garry acts as second reservoir, with a small concrete dam and tunnel leading to a 20 MW station at Invergarry, which discharges into Loch Oich.

The northern part of the scheme has two main reservoirs, Loch Loyne and Loch Cluanie, and I shall return to it later. Meanwhile, back to Telford's road, the C1144, which we rejoin at Tomdoun and follow westwards alongside Loch Garry towards Loch Quoich and Loch Hourn.

Approaching Loch Quoich, we notice its dam nestling comfortably in the landscape. Watch out for Telford's original roadway as it bears left just before the dam, stopped dead in its tracks by the towering structure. When the dam was built the loch increased in area from 3 to 7 square miles, and in depth by as much as 100 feet, and inundated a grand house. The present road, built in the 1950s around the artificial shoreline, rises to the right of the dam, keeping motorists dry and happy but consigning Telford's original roadway to the depths.

*A set of photographs taken in 2008 along the abandoned Telford road
between Tomdoun and the Cluanie Inn. Top left, leaving Tomdoun;
bottom right, approaching Cluanie; the other photos show the normally
flooded roadway and the bridges (Geograph/Trevor Wright).*

Quoich power station; and the approach to Loch Quoich dam (Geograph/Oliver Dixon).

For the next 6 miles the new road winds its way above the maximum fill level of the reservoir. The rise and fall of water level in response to rainfall and the demand for electricity generation is far greater than it was before the dam was built, and in the exceptionally dry spring and summer of 2010, long stretches of Telford's roadway were revealed.

About 4 miles beyond the dam, a three-span concrete bridge crosses a narrow finger of the loch, previously a dry glen. Two miles later the road leaves the lochside, merging once again with Telford's original route as it climbs gently to a watershed, beyond which several burns that previously headed west towards Loch Hourn now find their water piped to Loch Quoich for electricity generation.

Our road soon becomes altogether more adventurous – a single-track descent to sea level via a series of small bridges, blind hairpin bends, passing places, and steep gradients. But the challenge is rewarded by some magnificent scenery and the joy of arriving at a spectacular destination far from the madding crowd.

Loch Quoich: at typical level; and exceptionally low in 2010, revealing part of Telford's original roadway (Geograph/Jim Barton).

The final single-track descent towards Kinloch Hourn includes Telford bridges; and several blind bends (Geograph/Dave Fergusson).

Approaching Kinloch Hourn: 'Whatever brought you here?' (Geograph/Leo); and 'The end of the road' (Geograph/Dave Fergusson).

The settlement of Kinloch Hourn involves little more than a sprinkling of houses on the north shore of the loch, a short jetty or two, and a car park. More than any other place I know in Britain, it is truly 'the end of the road'. Keen walkers can follow a rough track 7 miles along the southern shore to beautiful Barrisdale Bay, a gateway to Knoydart, one of Scotland's most remote and challenging regions. If you are prepared to walk another 10 miles and climb 1,000 feet before dropping down to the village of Inverie on Loch Nevis, you will find yourself negotiating one of Scotland's 'coffin roads', an ancient track associated with death, loss, struggle, and the power of the church. For corpses were carried this way, sometimes in drenching rain or winter frosts, for burial on church land – a matter of authority as well as piety. And having arrived in Inverie (population about 100), you will either have to retrace your steps or take a boat to Mallaig: the hamlet, completely isolated from the national road

Left: The head of Loch Hourn (Geograph/Leo).

Right: Glen Arnisdale (Geograph/Julian Paren).

network, has won a Guinness National Record for remoteness. Its first-ever takeaway food was delivered in 2017 – by helicopter.

The first 5 miles of Loch Hourn are so narrow and hemmed in by mountains that it could easily pass for a Norwegian fjord. A road along the northern shore starts at Corran, runs past beautiful Glen Arnisdale, and continues west below the vertiginous slopes of a Munro, Ben Sgriol (unforgettable after a tough scramble to the summit on a wet and windy day). There are wonderful views across the loch to slightly higher Ladhar Bheinn, followed by a broad 10-mile northward sweep around a headland to the small village of Glenelg. The tidal strait of Kyle Rhea that separates the Isle of Skye from the mainland and was for centuries the

Loch Hourn from the summit of Ben Scriol (Wikipedia/Nick Ottery).

Looking across Loch Hourn to Ladhar Bheinn (Paul A. Lynn).

main crossing place for cattle as well as people, runs vigorously nearby. Nowadays it offers motorists an adventurous summertime alternative to Skye Bridge – a quaint vessel, said to be the world's only swinging car ferry.

Telford also explored the wild country between Loch Hourn and Glenelg in 1801–2, after convincing himself that Lochs Quoich, Garry, and Oich could provide the summit water supply for the Caledonian Canal. Of course he was no tourist; he had been tasked with assessing the suitability of various western sea lochs for fisheries, and of various tracks for upgrading into proper roads. Rather than return the way he had come, he decided to press on from Glenelg to Shiel Bridge and return to the Great Glen by a more northerly route. He would use much of it for the third of his Roads to the Isles, completed in 1813.

3 Invermoriston to Kyle of Lochalsh

See map on page 169

We will now explore the modern version of his northerly route ourselves – but in reverse order, from the Great Glen at Invermoriston (beside Loch Ness, about 6 miles north of Fort Augustus) to Kyle of Lochalsh opposite the Isle of Skye.

The first structure to claim our attention is Invermoriston Old Bridge, built over the River Moriston close to a modern road bridge and the spectacular River Moriston Falls. Generally

The old bridge at Invermoriston (Wikipedia/Jamesfcarter).

credited to Telford, it was started in 1805 but only finished eight years later thanks to 'idle workers' and a 'languid and inattentive contractor'. The two stone arches are supported by a large natural rock island in the middle of the river which simplified construction, supposedly reduced costs, and gave great strength in times of flood.

The first 15 miles of our journey along the A887 from Invermoriston to Bun Loyne follows the delightful River Moriston. At this stage it is a fine but little-used road, one of the safest trunk roads in the country. About 4 miles out of Invermoriston a sweeping bend takes us past the head of Dundreggan Reservoir – a sure sign that we are entering the northern territory of the Garry-Moriston Hydro-Electric Scheme.

The scheme's northern section has two main reservoirs (Loch Loyne and Loch Cluanie), several dams, and three power stations, rated at 15, 20, and 36 MW. The largest station, Glenmoriston, lies some 295 feet (90 metres) underground, beneath Dundreggan Dam. As we progress up Glen Moriston there is further evidence of the scheme, one of the largest in Scotland which, like the southern section along Glen Garry, became fully operational in 1962.

The River Moriston is soon crossed by two more bridges with strong Telford connections. The first, a few miles up the glen at Torgoyle, has a surprising history, for the bridge we cross today is a rare example of a replacement of a replacement of a Telford one that collapsed. And although I normally avoid lengthy quotes from official documents, in this case there is one too riveting to ignore: the 9th report of the Parliamentary Commission for Highland Roads and Bridges, dated April 1821,which illustrates only too clearly the destructive forces of nature that Telford and his assistants had to contend with in the West Highlands:

TORGOYLE BRIDGE

A serious casualty occurred in the beginning of the year 1818, on the 13th January, when an extraordinary flood took place in all the Rivers of the North of Scotland, and by it the Bridge of Three Arches [Telford's] over the River Morriston [*sic*], at Torgoyle, was swept away, not from any fault of the Masonry, which was excellent, but from a cause at once remote and unavoidable. A vast quantity of Birch Trees had been cut in a Forest above the Bridge, and deposited at such a distance from the Banks of the River, as to be deemed safe from the effect of any flood. But the elevation of the Water on the present occasion having been no less than Four feet perpendicular above any former example, the Timber was carried down to the Bridge, which after having been battered by about four thousand trees, was overwhelmed in such a manner that the destruction commenced with the upper part of the Arches, and the rest of the Timber (almost an equal quantity) demolished the Piers in its passage.

This was an emergency which called for all the activity of Mr. Mitchell, who received the intelligence at Dunkeld, on the 20th January, and was retarded in his return by the necessity of visiting the Laggan Bridge over the River Spey, which was endangered by the effects of the same flood. Having given directions for securing this important Bridge, he arrived at Inverness on the 25th January, after a dangerous journey through a snow storm, which covered the road in sundry places from six to twelve feet in depth. He forthwith hastened to Glen-morriston, and on his return to Inverness, on the 30th of January, transmitted a circumstantial Statement, which was answered by instructions to prepare materials and to construct a firm Timber Bridge, at Torgoyle, as soon as the state of the River permitted; the injured Bridge at Drumnadrochet [*sic*] having been already rendered passable, on the 5th February, in consequence of directions given by Mr. Mitchell for a temporary repair.

Materials were accordingly collected, and a hard frost having reduced the quantity of water in the River Morriston, Mr. Mitchell went to Torgoyle on the 25th February, and in the course of the next four days, constructed six piers, formed of four piles each. These were driven into the Bed of the River, by means of a piling engine, which with its supporting scaffold, was to be moved for every Pile: an operation which kept the workmen for hours together in the River, at that time four feet deep. This was no small effort in a sharp frost; but there was no alternative, and the Timber Bridge was rendered passable in the beginning of March, having been completed at an expense of £167.

Leeft: Torgoyle Bridge, completed by Joseph Mitchell in 1823 (Geograph/John Allan).

Right: The main arch of the old bridge at Ceannacroc,
completed by Telford in 1811 (Geograph/Jim Barton).

The hero in this case was not Thomas Telford, who had built the original three-arch bridge with excellent masonry in 1811, but Joseph Mitchell, son of Telford's 'Tartar', John Mitchell; Joseph was acting as principal inspector in the West Highlands and had been given the unenviable task of constructing a temporary timber bridge after the disastrous 1818 flood. By 1823 he had also built the fine stone replacement we see today, 'a large 3-span bridge, the centre arch being slightly wider and higher; tooled and pinned rubble with tooled ashlar dressings; three segmental-headed tooled ashlar rings, the dressings being alternate blocks of dark schist and lighter granite, springing from rusticated masonry cutwater abutments'. It was awarded a Category A listing in 1971. Full marks to Joseph Mitchell.

The second notable bridge, a Telford original of 1811, is at Ceannacroc, about a mile short of Bun Loyne. Built with two spans, a 50-foot main arch and a 36-foot flood relief arch, it survived the flood that destroyed Torgoyle Bridge. Still standing now, but bypassed by a later bridge, it is approachable by a very minor right turn off the A887.

There is an important motoring event at Bun Loyne. The A887 on which we have been travelling is joined from the left by the modern A87, coming over the hills from Glen Garry and carrying – in the summer months at least – a great deal of holiday traffic on its way to and from the Isle of Skye. So thanks to the Garry-Moriston Hydro-Electric Scheme, which caused considerable rerouting of roads in the late 1950s, our uncluttered A887 now forfeits its identity to the A87 as we continue westwards towards Kyle of Lochalsh.

Whatever one thinks of hydro-electric schemes and lochs expanded into reservoirs, it is hard to fault their green credentials: electricity generation without carbon emissions, and a free and eternal supply of 'fuel'. But I confess to feeling ambiguous about dams inserted into beautiful landscapes, especially when they are unanticipated: the greatly increased rise and fall of shorelines tends to unsettle me. However there may be a compensation: a hydro-electric reservoir, far greater than the original river or loch it has overwhelmed, sometimes

The westward panorama over Loch Loyne (Paul A. Lynn).

creates its own beauty in mountainous country by adding a majestic aqueous component to the mix. Nowhere have I felt this more keenly than among the hills between Bun Loyne and Glen Garry, gazing astonished at multiple Munros and a westward panorama over the greatly expanded Loch Loyne. It was easy for me to forgive the dam which had made it possible.

Cluanie Dam is larger than the one across Loch Loyne, with a length of 2,215 feet (675 metres) and height of 131 feet (40 metres), and is certainly impressive as we approach from Bun Loyne. The route of the old road may be seen below to the left, snaking towards the high wall which has effectively elevated and expanded Loch Cluanie into a reservoir to feed Ceannacroc power station nearby. As we pass along the north side of the loch there are occasional glimpses of an old road to one side or the other, reminding us that Telford came this way – as had Major Caulfeild, General Wade's successor as Scotland's builder of military roads, half a century earlier.

After 8 miles alongside Loch Cluanie we come to the Cluanie Inn, well known to generations of climbers and hill walkers. Shortly afterwards we cross the watershed and begin a steady 1,000-foot descent through Glen Shiel to Loch Duich on the western seaboard.

There are treats in store as we negotiate Glen Shiel, guarded by multiple Munros including the Five Sisters of Kintail, which rise to a maximum of 3,505 feet (1,068 metres) above sea level. I remember a superb day's hill walking which, after a soggy crossing of marshland

Cluanie Dam (Geograph/James Allan); and the Cluanie Inn (Geograph/John Allan).

In Glen Shiel (Geograph/Jennifer Jones).

beside the River Shiel, gave us wonderful views from the summit of the second highest sister, and the greatest respect for a shepherd and his dog rounding up sheep near the summit of the first.

The small settlement of Shiel Bridge at the lower end of the glen is the meeting point of the A87 and a minor road leading 9 miles over the spectacular Mam Ratagan Pass to Glenelg and the Kylerhea ferry service to Skye (capacity six cars and, at the time of writing, summer season only). In the early 1800s this road, which had acted as a drove road for centuries, was still the main route to Skye. Assuming we have time to spare, and a little of the adventurous spirit, let's take it – and on the way learn more about its history.

Turning left off the main road we cross a Telford bridge over the River Shiel. Few people realise that much of the route to Glenelg and the ferry was upgraded from drove road to military road by Major Caulfeild (successor to General Wade) in 1755; that Dr Samuel Johnson, giant of English letters and creator of the first English dictionary, travelled along it in great discomfort with his biographer James Boswell on their famous tour of the Hebrides in 1783; and that Telford came this way after surveying Lochs Garry, Quoich, and Hourn in 1801. So we are now following some very famous people – fortunately on a surface that owes much to the 1980s, when it was substantially improved for winter conditions and modern traffic levels.

But the road is still narrow, often single track, and well-endowed with sharp bends. After skirting the flat southern shoreline of Loch Duich we begin a long and tortuous climb to the 1,116-foot (340-metre) summit of the Mam Ratagan Pass. Much is hemmed in by trees, but luckily there are viewpoints about two thirds of the way up and at the top, giving wonderful panoramas over Loch Duich to the Five Sisters of Kintail. Beyond the summit the road seems at times to lose its sense of direction – and height – but it eventually settles into a gentle descent along a delightful valley towards Glenelg village and Bernera Barracks. A right turn shortly before the village leads about 3 miles to the historic swing ferry across the narrow strait of Kyle Rhea to the Isle of Skye.

The Bernera Barracks were built a few years after the Jacobite rebellion of 1715 by a British government anxious to dissuade the Highlanders from further military adventures – rather unsuccessfully, as it turned out, thanks to Bonnie Prince Charlie. By 1755 Caulfeild had upgraded existing tracks into a military road that ran all the way from the Great Glen to the

Left: Telford's bridge at Shiel Bridge (Geograph/John Ferguson).

Right: View from the Mam Ratagan Pass towards the head of Loch Duich, the settlement of Shiel Bridge, and the Five Sisters of Kintail (Geograph/Russel Wills).

Left: Bernera Barracks (Wikipedia/Pasicles).

Right: The swing ferry across Kyle Rhea to Skye (Wikipedia/Wojsyl).

barracks and Kyle Rhea. He headed west from Fort Augustus through Inchnacardoch Forest and climbed to a height of 1,300 feet (~400 metres) before dropping down to Glen Moriston, not far short of Bun Loyne, then continued slightly to the north of the present A887 and A87 before descending to the coast at Shiel Bridge and tackling the Mam Ratagan Pass.

Today the derelict barracks are protected as a scheduled monument, and present a grim reminder of troubled times: the troops garrisoned here were expected to control the use of the road and the crossing to Skye during the unpredictable Jacobite years, and were finally withdrawn in 1797, four years before Telford arrived on the scene.

Tempting though it is to take the Kylerhea ferry 'over the sea to Skye', I suggest deferring the pleasure for the time being and returning over Mam Ratagan Pass to Shiel Bridge. From there we will follow the A87 to Kyle of Lochalsh, continuing the main line of Telford's third Road to the Isles. The 18-mile run will reintroduce us to the magic of West Highland sea lochs and Eilean Donan, one of Scotland's most celebrated castles.

Eilean Donan, perched on a small island in Loch Duich, was built in the 13th century to protect the lands of Kintail from the Viking raiders who had for over 400 years terrorised much of the north of Scotland and the Western Isles. The following centuries saw the castle playing an important part in the volatile and often violent history of the Highlands, culminating in 1719 with an occupation by Spanish soldiers awaiting the arrival of weapons and cannon from Catholic Spain. The London government caught wind of the intended uprising and sent three heavily armed frigates to bombard the castle walls. Following the Spaniards' surrender, 343 barrels of gunpowder were discovered and used to blow up what remained. The stark ruins of Eilean Donan then lay open to the elements for about 200 years, until the magnificently named Colonel John Macrae-Gilstrap bought and restored it to its former splendour. Re-opened in 1932, it has since become one of Scotland's major tourist attractions.

One glance at a map shows why Eilean Donan proved a powerful defensive asset: it is located at the confluence of three sea lochs: Loch Duich; Loch Long, which stretches 5 sinuous miles inland among the hills and may have provided a safe base for a home fleet;

Eilean Donan Castle (Geograph/Gordon Hatton).

and Loch Alsh, with just two narrow exits towards the open sea, at Kyle of Lochalsh and Kyle Rhea. As we pass along the A87 on Telford's 18-mile stretch from Shiel Bridge to Kyle of Lochalsh, enjoying superb views towards the mountains of Skye, we might reflect on the strategic location of Eilean Donan throughout the centuries when west coast warfare was conducted more at sea than on land.

I first arrived in Kyle of Lochalsh many years ago in a little old Morris 8 which had been mothballed during World War II for lack of petrol and rescuscitated in the peace that followed. Meeting up with three friends who had travelled by train, we crossed over to

Kyle of Lochalsh (Geograph/David Purchase).

Kyleakin on Skye on one of the old ferries run by Caledonian MacBrayne. Romantically slow but increasingly overwhelmed by summertime tourist traffic, this particular ferry was finally put out of business in 1995 by the superb Skye Bridge. Little did we realise at the time that Thomas Telford had built his third Road to the Isles this way, and had pioneered the road network we were about to enjoy on the magical Isle of Skye, or that he had extended his Kyle of Lochalsh road, from Auchtertyre, 7 miles north to Strome Ferry on Loch Carron, where it would link up with the road he was driving east from Inverness to Strome Ferry by a more northerly route.

4 Inverness to Strome Ferry

See map on page 169

To describe another of Telford's routes as a Road to the Isles is, I admit, a little unusual. The first 20 miles or so, from Inverness via Beauly and Muir of Ord to Dingwall, are not obviously aimed at the west coast; it is only after travelling through Contin and Garve on the A835, turning left onto the A832 towards Achanalt and Achnasheen, and left again onto the A890 towards Lochcarron, that we are definitely heading for the coast opposite the Isle of Skye.

Actually we have travelled this way before, in the company of Telford, Rickman, and Southey. In the West Coast Excursion section, I discussed the poet's account of their journey along the partly built road from Garve to Lochcarron in September 1819. Southey set the scene:

> Left the Ladies and the children at Dingwall; they were to return in the Coach to Inverness and there wait for us. We with a chaise and Mitchell's gig set off at six, to cross the island by the Loch Carron Road, from sea to sea. The chaise horses had been sent off yesterday, one stage to Garve, and we took a pair of the Coach horses so far, which enabled us to perform the whole journey in one day.

Lochcarron village (Geograph/Jim Barton).

You may like to refer back to the West Coast Excursion section, leaving me to cover the 14-mile northward extension of Telford's road from Lochcarron to Shieldaig on Loch Torridon – an extension completed by 1819 but apparently not seen by Robert Southey. Today it forms the southernmost section of the A896.

The village of Shieldaig and the Applecross peninsula were historically isolated. Prior to 1800 they had relied on rough tracks and sailing boats for all contact with the outside world. Telford must have seen Shieldaig as a prime candidate for a fishing port – provided it was given a decent road connection. The route he chose was straightforward, without the challenging gradients or hairpin bends of the Mam Ratagan Pass.

However, we have not quite finished with mountain passes. Five miles after leaving Lochcarron for Shieldaig, and just beyond the head of Loch Kishorn, we notice a left turn onto a minor road that leads 11 miles over the mountains to Applecross village on the western seaboard. This is the famous drover's road, *Bealach na Bà* (Pass of the Cattle), built not by Telford but by MacKenzie of Applecross at his own expense in the early 1820s. The third-highest road in Britain, single-track with passing places, it rises more than 2,000 feet (610 metres) above sea level and zig-zags crazily up steep gradients. I remember the days when the roadway seemed as much grass as bitumen, with a sprinkling of gravel on the summit. Even today it is regarded as something of a motoring adventure, and there are warning signs at the roadside: 'Not advised for learner drivers, very large vehicles or caravans after first mile', and 'Road normally impassable in wintry conditions'. Perhaps we can chicken out, plead lack of time, continue our journey along the A896 to Shieldaig, and reserve tomorrow for the delightful 25-mile alternative: a low-level coast road (C1091) built between 1964 and 1976 that winds from Shieldaig along the shore of Loch Torridon to the nothern tip of the Applecross peninsula, then down the coast to the eponymous settlement. Along the way we will, with luck, enjoy superlative views across the Inner Sound towards the mountains of Skye.

Left: Bealach na Bà in winter (Geograph/Alan Reid).

Right: The mountains of Skye seen from Applecross (Geograph/Lisa Jarvis).

Left: Shieldaig (Wikipedia/DeFacto).

Right: A train at Stromeferry Station on the Dingwall to Kyle of Lochalsh line (Geograph/Trevor Littlewood).

Telford was surely right about Shieldaig's potential as a fishing port, because its name comes from the Old Norse for 'herring bay'. Actually the village had been founded a couple of years before Telford arrived, to train Highlanders as sailors for the British navy; but following the Battle of Trafalgar in 1805 and Napoleon's exile on Elba, fishing looked the better option. Today Shieldaig's small community has its own school, church, village hall, and smokehouse – plus B&Bs and holiday houses. It still attracts fishermen – and, now that the A896 has been extended eastwards towards the wonderful mountains of Torridon, hill walkers.

Returning south to Lochcarron, you may recall that when Telford, Rickman, and Southey reached the village from Inverness in 1819, they intended to return to the Great Glen by

Loch Carron (Geograph/Peter Jeffery).

the Strome Ferry across Loch Carron, then along Telford's recently built section of road to meet what is now the A87 between Kyle of Lochalsh and Shiel Bridge. But the new ferry boat designed to take horses and carriages, promised by the Laird of Applecross, was nowhere to be seen, so they returned the way they had come.

In those days Strome Ferry was essential because there was no road or railway along the southeastern shore of Loch Carron. The railway, originally known as the Dingwall and Skye Line, arrived in 1870; and the road, an extension of the A890 from Achnasheen, far more recently. They sounded the death knell for the historic ferry, but its memory lives on in the name of the railway station, Stromeferry – spelled as one word rather than two.

The Inner Hebrides

Skye

Back in 1819 Telford, Southey, and Rickman had caught sight of the mountains of Skye from a hillside near Strome Ferry; but their unavoidably hit-and-miss transport arrangements prevented them from reaching the magical island. We, too, have had tantalising glimpses of Skye from the mainland – and fortunately we can indulge them. It is time to go 'over the sea to Skye' in the steps of Telford, who did so much to pioneer the island's road network.

I suggest we take the historic ferry from Glenelg to Kylerhea, as Telford did. For centuries the narrow tidal strait of Kyle Rhea was the main crossing point to and from Skye, for animals as well as humans. Cattle reared on Skye and other Hebridean islands were driven across the strait on their way to mainland markets; but the drovers dared not force weary animals through its dangerous currents, and were easily tempted by a meal and a nap at the nearby inn. Then, at or near slack water, they funnelled the cattle down a narrow passageway between rocks for an unpremeditated 600-yard swim. As we board today's community-owned, six-vehicle car ferry, full of admiration for a crew who can swing a turntable without motor assistance, perhaps we should spare a thought for the drovers of old.

Our own 600-yard crossing ends at Kylerhea, the Skye settlement closest to the mainland. It is well known for an otter haven and other wildlife including seals, golden eagles, and the occasional white-tailed sea eagle. It is also the starting point for a delightful 7-mile drive inland along a well-maintained single track road with passing places and ancient bridges, originally engineered by Thomas Telford. Outside the main holiday season we will probably have it to ourselves, a rare motoring treat in the 21st century; in season, we are more likely to meet a few cars hurrying towards the ferry. The road strikes north between hills to meet the A87 about 4 miles east of Broadford.

Before the Skye Bridge was opened in 1995, the island's main trunk road through Broadford and Portree to Uig was designated the A850/A856. But it has been renamed the A87, a logical extension of the A87 that runs from the Great Glen to Kyle of Lochalsh. It is

Over the sea to Skye on the Kylerhea swing ferry (Geograph/Craig Wallace).

as though Telford always envisaged one of his Roads to the Isles ending at a port in the far northwest of Skye, from where boats would sail for the Western Isles. Of course at that stage a bridge from the mainland to Skye could have been little more than the engineer's pipedream.

On reaching the A87 we turn left towards Broadford, leaving Skye Bridge for another day. At this point I should make a confession: although we will travel all the way to Uig on the A87, I wish to inject a few diversions and anecdotes. For ever since my first visit to Skye, the largest of the Inner Hebridean islands, I have been fascinated by its wonderful landscapes, seascapes, and mountains.

The first diversion starts almost immediately. A mile or two short of Broadford we turn left onto another route pioneered by Thomas Telford: the A851, which runs 16 miles south across the Sleat peninsula to the village of Armadale. Large Calmac ferries run a regular service from Mallaig to Armadale – a 5-mile crossing rather than the mere 600 yards from Glenelg to Kylerhea. Many visitors to Skye come this way, not realising that the A851 is, in effect, an extension of Telford's best-known Road to the Isles, the A830 from Fort William to Mallaig.

Calmac ferry Coruisk *approaches Armadale from Mallaig (Geograph/M.J. Richardson).*

The Sleat peninsula, traditionally rather isolated from the rest of Skye, has maintained its Gaelic traditions. As recently as 2011 almost 40 per cent of its population were recorded as Gaelic speakers, and the local primary school is far more Gaelic than English. Sleat, often called 'the garden of Skye', is gentle country by Skye standards, flattish and fertile rather than mountainous.

But for centuries Sleat's history was anything but gentle. The former clan chiefs, MacDonalds of Sleat, left a sad and muddled legacy that is difficult for an outsider to grasp, let alone summarise. From the 16th century onwards it seems to have involved (in rough chronological order) murderous infighting; vicious feuds with Skye neighbours, especially the MacLeods of Dunvegan; defeat by parliamentary forces in the civil war; attempts to sell tenants into slavery in the American colonies; support for the first Jacobite rebellion (but not the second); a grim catalogue of 19th-century clearances, emigration, and destitution; and a final alienation of clan families from their chiefs, who vamoosed from the south of Skye to the north of England. Armadale Castle, built in the early to mid-19th century, was finally abandoned by the MacDonalds in 1925 and is now a picturesque ruin. But the estate has been taken over by a charity, the Clan Donald Lands Trust, and rebranded as a modern visitor attraction:

> Welcome to our magnificent Highland estate, the spiritual home of Clan Donald, on the magical island of Skye. Explore the historic gardens and woodland trails around the romantic ruins of Armadale Castle. Discover 1,500 years of Highland history in our award-winning Museum of the Isles.

I confess to feeling slightly confused.

Returning along the A851 to its junction with the A87, we turn left once again to the centre of Broadford, from where I cannot resist another diversion – even though it is along a minor road which, to the best of my knowledge, has nothing to do with Thomas Telford. The B8083 winds 14 miles in a generally southwesterly direction to the coastal village of Elgol. Skye mists permitting, it will lead us towards an unforgettable seascape and the island's iconic mountains, the Black Cuillins.

For the first 6 miles or so we skirt around Beinn na Caillich (2,403 feet/ 732 metres), one of the Red Hills. Although they lack the high drama of their more celebrated neighbours, they

Left: Armadale Castle, Sleat (Geograph/Martyn Gorman).

Right: Waiting for the ferry at Armadale in 1976 (Paul A. Lynn).

have their own gentler beauty: composed mainly of a pinkish granite eroded into smooth, rounded hills flanked by long scree slopes, they are famed for glowing warmly in morning or evening sunlight.

We soon come to Torrin, a village beside Loch Slapin. Ahead rises the unmistakable outline of Bla Bheinn (3,042 feet/ 927 metres), a double-peaked Munro and surely one of Scotland's finest mountains. It signifies wild country ahead – a land of high summits, gushing burns and, in sunshine and showers, a dramatic colour palette. Bla Bheinn has much of the character of the main Cuillin ridge which, at this point, it obscures from view. To see that, we must round the head of Loch Slapin and continue south to Elgol.

Of all the seascapes I have enjoyed in the West Highlands, none surpasses that from Elgol's shore towards the Black Cuillins. But I must add a caveat: to avoid disappointment, choose a clear day with a visibility of at least 8 miles. And if you have time for a boat cruise across Loch Scavaig to the shore beneath the Cuillin ridge, you will discover one of Scotland's – I dare say Europe's – most stunning freshwater lakes, cradled within a mountain amphitheatre. Loch Coruisk is a prime example of what the Victorians called *sublime,* nature at its most majestic and awe-inspiring.

Returning to Broadford the way we came (there is no other road), we continue west along the A87 towards Sligachan. The scenery along this coastal stretch is dramatically confusing, with the Red Hills vying for attention inland and the islands of Scalpay and Raasay appearing intermittently offshore. At the head of Loch Ainort the road strikes inland to skirt around Glamaig (2,537 feet/ 773 metres), the highest of the Red Hills. An adventurous alternative is to continue along the coast through the hamlet of Moll, and enjoy a primitive single-track road that avoids 99.9 per cent of the traffic scurrying along the A87.

Sligachan is one of the best-known meeting places on Skye, not least for the annual hill race up and down the slopes of Glamaig, which starts and ends at the Sligachan Hotel. Sgurr nan Gillean (3,167 feet/ 965 metres), the first of 12 Munros forming the 9-mile Cuillin ridge, rises majestically to tempt mountaineers; another Red Hill, Marsco (2,414 feet/ 736 metres), shows its distinctive profile. We are offered solid evidence that Telford came

High peaks, a gushing burn, dramatic colours (Paul A. Lynn).

*A magnificent seascape: the view from Elgol across Loch Scavaig
towards the Black Cuillins (Geograph/Andrew Hill).*

A Victorian view of Loch Coruisk, painted in 1874 by Sidney Richard Percy (Public Domain).

*Above: Telford's bridge at Sligachan.
Ahead stretches Glen Sligachan, with the
well-known outline of Marsco at centre
left (Geograph/Chris Downer).*

*Right: Loch Coruisk and the Cuillin ridge
seen from Sgurr na Stri (Paul A. Lynn).*

this way by Sligachan's old bridge, a three-span rubble structure with triangular cutwaters which he built in the 1810s. It makes a fine sight as we cross the modern bridge over the River Sligachan.

All this reminds me of the day I arrived at Sligachan with my three student friends and a determination to see Loch Coruisk. We trekked up the glen past Marsco on what is nowadays advertised as the Sgurr na Stri Trail. It turned out to be 19 miles return – rather more than expected – and our snacks of chocolate, mint cake, oranges, and bottled water were thoroughly depleted by the time we regained the Sligachan Hotel and its bar with our tongues hanging out. But the day's effort had rewarded us with an unforgettable experience – the wonderful view of Coruisk and the Cuillin ridge from the upper slopes of Sgurr na Stri.

The dramatic and occasionally hazardous Cuillin ridge soars above steep cliffs and long scree slopes. Sgurr Alasdair (3,255 ft/ 992 metres) is the highest of 12 Munros, and the Inaccessible Pinnacle perched atop Sgurr Dearg (3,234 ft/ 986 metres) the most revered. They are composed of two types of volcanic rock: gabbro, dark and coarse-grained, with a superb grip for the mountaineer; and basalt, pink, fine-grained, and dangerous when wet. Many people are surprised to learn that both are comparatively young by geological standards – a mere 60 million years compared with the 2 to 3 billion years of the Lewisian gneiss which outcrops in the Western Isles and northwest Scotland, including the Sleat peninsula on Skye.

*Above: Talisker Distillery at Carbost
(Geograph/Richard Dorrell).*

*Left: The Inaccessible Pinnacle
(Geograph/Ian Taylor).*

On the far side of Sligachan bridge there is a left turn off the A87 onto the A863, signposted to Dunvegan; and we will take it, because it was originally a Telford road. It will lead us about 20 miles through delightful coastal scenery to one of Skye's most visited buildings, Dunvegan Castle, ancestral home of the MacLeod of MacLeod.

As we have seen, Telford's work in Scotland involved him in all sorts of encounters with Scots grand and humble, and he felt at ease in a wide variety of professsional and social situations. But although he was working for, and fully backed by, the London government, he had to watch his step. Skye's turbulent history included long-running feuds and rivalries between the MacLeods and the MacDonalds; a dangerously close connection with Bonnie Prince Charlie's 1745 rebellion; and a gathering storm of clearances. New infrastructure – roads, bridges, harbours – and the social upheaval they implied were often welcomed by the powerful, but distrusted by their tenants. And on Skye Telford had to balance the interests of two rival clans: a new road from Sligachan to Dunvegan for Clan MacLeod presumably went hand in hand with the one from Broadford to Armadale for Clan MacDonald.

About 4 miles along the road to Dunvegan a minor road, the B8009, offers a detour to Carbost, which may well appeal if you are a lover of the wee dram. For this is the home of the famous Talisker, Skye's oldest working distillery, set on the shores of Loch Harport, with dramatic views of the Cuillins.

Continuing along the A863 we soon pass the villages of Bracadale and Struan, and see a small lane on the left signposted to Ullinish and one of the most historic buildings on Skye

– Ullinish Country Lodge (previously known as the Ullinish Lodge Hotel), nestling in the landscape close to the shores of Loch Bracadale.

I have vivid and happy memories of the hotel because it was the first place where I and my friends ever stayed on Skye. We had booked in for a week, and although that may seem rather an extravagance for mere students the hotel was far less impressive than it is today – and, in mid-September, far less busy. Each of us paid one guinea (£1.05) per day for dinner, bed, breakfast, and a picnic lunch prepared by redoubtable Mrs Clark, the large and warm-hearted hotelier. A special memory is the way she cooked magnificent scotch pancakes on her stove for us and, the weather being equally magnificent for the whole week, conveyed them to us through the kitchen window to enjoy outside at a table set on the front driveway.

But in those far-off days I knew next to nothing about the history of the Ullinish Lodge Hotel, except that it figured prominently on my Bartholomews half-inch map of Skye. I would have judged it a Victorian building, but it was actually built in 1757, the year of Thomas Telford's birth. The celebrated Dr Johnson and his biographer James Boswell stayed there in the autumn of 1773 while touring the Western Isles, and Boswell recorded 'We got to Ulinish [*sic*] about six o'clock, and found a very good farm-house, of two stories … There is a plentiful garden at Ulinish (a great rarity in Sky) [*sic*] and several trees.' The wonderful view of the Cuillin ridge seen from Ullinish reminded him of mountains in Corsica. That has certainly not changed; nor has the superb coastal scenery around lochs Harport and Bracadale.

Returning to the main road, we complete the 10 miles to Dunvegan Castle and find it perched dramatically above an inlet of Dunvegan sea loch. It was first built in the 13th century and, after a long history of modification and repair, was completely rebuilt in the 19th century in a strange mixture of styles. Now a popular tourist destination, it is notable for heirlooms collected by the MacLeod family over many hundreds of years. Outside the castle walls are 5 acres of superb gardens, and a vast estate of 42,000 acres which, for good measure, includes the Cuillin mountains.

Left: Ullinish Country Lodge, with the Cuillin ridge in the background (photo: Stella Morris).

Right: Oronsay Island on Loch Bracadale (Geograph/Ian Taylor).

Dunvegan Castle (Geograph/John Lord).

Our next major aiming point is Portree, Skye's capital. Rather than return to Sligachan along the A863, we will take the A850 – another road pioneered by Telford – out of Dunvegan village. It gives me an excuse for another diversion along a minor road strongly associated with Telford and very much associated with Skye's social history.

The B886 leaves the A850 about 3 miles from Dunvegan, signposted to the Waternish peninsula. Almost immediately we notice the Fairy Bridge, where the Chief of the MacLeods is said to have been presented (presumably by fairies) with the 'magical' flag now held in Dunvegan Castle. The road continues 4 miles to Stein, where in 1786 the British Fisheries Society decided to develop one of its fishing ports, to be designed by Telford as part of his ongoing commitment to the society. However it seems that a lack of local enthusiasm had

A Telford terrace at Stein includes the Stein Inn (Geograph/Anthony O'Neil).

halted the scheme by 1796, and when Telford returned to Skye in the early 1800s his mind was focused on major projects for the British government. Today, little Stein beckons us with a fine terrace of whitewashed Telford buildings overlooking a bay furnished with a modest jetty. The delightful Stein Inn, said to be the oldest on Skye, proudly proclaims its date – 1790.

I am unsure how far Telford's original road continued beyond Stein, but we will follow the present one 3 miles north to the small settlement of Trumpan. The route and coastal scenery are delightful, and deliciously uncrowded, even in the summer months.

Sadly, however, the distant history of Trumpan is anything but delightful. The year 1578 witnessed an incident which, even by the standards of 16th century clan feuding, plumbed the depths of brutality. Eight boats of the Uist (Western Isles) branch of Clan MacDonald sailed to Waternish and landed at Ardmore Bay near Trumpan. Shrouded by sea mist, their crews crept up on Trumpan church, set fire to the thatched roof, and burned alive all the MacLeod worshipers – apart from a young girl who escaped through a window and raised the alarm. MacLeods from the wider area gathered and followed the MacDonalds back to their boats, which had been left stranded by a retreating tide. At this point the MacLeods are said to have unfurled their famed Fairy Flag and killed the entire raiding party in a violent skirmish.

Unsurpringly, the MacDonald raid was itself retribution for a previous MacLeod outrage. A raiding party of MacLeods had landed on the island of Eigg (about 10 miles south of Skye) the previous year and found the resident MacDonalds hiding in a cave. A fire was lit at the entrance and 395 islanders suffocated. Not surprisingly their hiding place is known as Massacre Cave. It can still be visited.

There is something to be said for getting all the bad bits over and done with, so I will mention one more lamentable episode in Trumpan's troubled history. It concerns a privileged couple intent on mutual destruction, and the story starts in Edinburgh, not Waternish.

Lord and Lady Grange lived in Edinburgh until their marriage broke down in 1730. She moved out, citing her husband's extramarital affairs while away in London as an MP, and took to standing outside his house shouting obscenities and threatening to reveal his Jacobite sympathies. This was too much for Lord Grange, who had her kidnapped, shunted incommunicado around the Highlands and Islands, held prisoner on remote St Kilda for eight years, and finally moved to Skye, where she was left alone in a cave. The poor wretch died three years later at Trumpan. In the meantime Lord Grange had announced her death and arranged a fake funeral in Edinburgh.

It is hardly surprising that Trumpan, once a thriving medieval township, failed to re-cover from its troubles. The burnt-out church was abandoned, but the graveyard continued in use. Among many tombstones visible today is one for Lady Grange. I have been struck by the contrast between violent stories of man's inhumanity to man (and woman) and the wonderful serenity of the Waternish peninsula – especially when on a summer's day a gentle breeze wafts across the Lower Minch from the Western Isles and breathes new life into the hearts of a welcoming community.

Telford must surely have been aware of the local history when he selected sites for his Highland churches in the 1820s. It is easy to see why one of them was at Hallin, just 2 miles

*The churchyard at Trumpan, with the ruined church in the background
(Geograph/Richard Dorrell); and Ardmore Bay, with the Western Isles
just visible on the far horizon (Geograph/Uilleam Donnachaidh).*

from the burnt-out shell at Trumpan. Now, no longer a parish church of the Church of
Scotland, it has been converted to private use.

We now head back through Stein to the A850, which takes us about 17 miles eastwards to
a junction with Skye's main trunk road, the A87. Here we turn right to Portree. Unlike little
Stein, where plans for a proper fishing port floundered badly, Portree's went ahead under
Telford's supervision. He could hardly have guessed that his fine quayside would become a
major departure point for emigrants, especially to Canada, after his death. Today it is at the
heart of a flourishing town, backed by brightly coloured buildings and well used by fishing
boats, pleasure craft, and an all-weather lifeboat. Portree has spread out in recent years with
new housing, supermarkets, and a Heritage Centre that brings Skye history to life for the
town's many visitors.

Left: The former Telford church at Hallin, Waternish (Geograph/Dave Fergusson).

Right: Portree Harbour, designed by Thomas Telford (Wikipedia/DeFacto).

Apart from the famous old bridge at Sligachan and the harbour at Portree, there is rather little evidence of Telford's original work on Skye compared with other locations we have visited – especially along his Roads to the Isles. I have sometimes used a good roadmap to search for rivers he must have crossed, then investigated the sites for evidence of ancient bridges. A good example comes after we leave Portree and head north along the A87 towards Uig. The map shows three substantial rivers ahead: Haultin, Romesdal, and Hinnisdal. The first two provide no such evidence – but glance upstream from Hinnisdal's modern bridge and you will see the abutments of a ruined single-arch one – Telford's, almost certainly.

Uig, 16 miles from Portree, is an attractive fishing and crofting village in the Trotternish peninsula, spread generously around one of Skye's largest and most sheltered bays. It is principally known to the outside world for ferry services to and from the Western Isles, including Tarbert on the Isle of Harris and Lochmaddy on North Uist. The King Edward Pier, extended in 1894 at a cost of £9,000, was officially opened in 1902 by King Edward VII and Queen Alexandra, and still acts as a focal point for fishing activities.

Trotternish, underlain by basalt, has richer soils than most of Skye, and some spectacular rock features. I once took the unclassified single-track road that leaves the A855 coastal route a mile or so beyond Uig, signposted to the Quirang, and discovered some of the most remarkable landscapes in Scotland. After climbing steadily up to the Trotternish ridge, the road dives crazily down on the eastern side towards Staffin. The drama is thanks to a massive ancient landslip that created high cliffs, hidden plateaux and soaring rock pinnacles. If you return to Portree via Staffin and the A855, keep a lookout for the Storr Pinnacles and the Old Man of Storr, some of the most iconic rock clusters in Scotland.

It is almost time to say farewell to Skye and its Telford connections. We will leave Portree as we arrived – along the A87 through Broadford. But rather than take Telford's minor road to the Kylerhea ferry, we will continue 4 miles to the elegant Skye Bridge, which since 1995 has

King Edward Pier, Uig
(Geograph/Alan Reid).

Walking in the Quirang (Geograph/Malcolm Neal).

The Storr Pinnacles (Geograph/John Allan).

*Farewell to Skye: looking back from Kyle of Lochalsh to the Skye Bridge,
with the Red Hills and Cuillins beyond (Geograph/Jim Barton).*

greatly simplified crossings to and from the mainland. Not without controversy in its early
days, the bridge upset certain locals who had viewed their ferry as the only valid way 'over
the sea to Skye', and resented the high toll charges (which remained in place until 2004). But
Telford, the master bridgebuilder and roadmaker who did so much to ease Skye into the 19th
century, would surely have applauded Skye Bridge as a much-needed addition to the island's
civil infrastructure.

Mull and Iona

Mull is the second-largest island of the Inner Hebrides. It is also our next destination – not
for its roads, which were little influenced by Thomas Telford, but for his excellent harbour at
Tobermory, a welcome sight to sailors seeking refuge from the North Atlantic.

Tobermory Bay bay first came to national attention in 1588, following the Spanish
Armada's defeat in the English Channel at the hands of Sir Francis Drake. Many Spanish
ships attempted to flee up the North Sea and around the coasts of Scotland, but most were
severely damaged or sunk by violent storms. One ship, loaded with gold and silver intended
to bankroll the invasion of England, is believed to have ended up at the bottom of Tobermory
Bay. Exactly how it would have got there is unknown. There was no significant settlement
around the bay at the time, nor even when the Royal Navy used the anchorage during the
Jacobite rising of 1745; but by 1788 the British Fisheries Society was planning a fishing port
there, to be known as British Harbour. Telford's long association with the society produced
some free advice, resulting in the fine quayside of 1814 known as Fisherman's Pier.

Tobermory Bay is nowadays used by cruise ships and diving groups as well as fishing
boats, yachts, and a lifeboat. The Harbour Association, founded in 1983 as a not-for-profit
organisation, manages the bay's facilities on behalf of visitors and the community. Tobermory
town is an attractive destination in its own right, with a good range of restaurants, bars, and

Left: *Sailing to the Isle of Mull. Ben More, its only Munro, is just visible in the distance (Geograph/Donald MacDonald).*

Right: *Tobermory Bay and Harbour (Geograph/Clive Perrin).*

shops that cater for most needs including chandlery. Mull Aquarium is Europe's first catch-and-release aquarium, with a seasonal display of magnificent local marine life, thoughtfully returned to the sea every couple of weeks. You may also visit Mull's whisky distillery, museum, arts centre, and theatre; stroll in beautiful Aros Park; or walk to the lighthouse built in 1857 by famous civil engineers David and Thomas Stevenson.

Although Telford's contributions to Mull are largely confined to Tobermory, the island has a great deal more to offer – so much so that I hardly know where to start. On balance I suggest driving down the A848 and A849 to Craignure (from where ferries sail to Oban), continuing along the A849 amid some of Mull's most spectacular mountain scenery, and finally skirting alongside Loch Scridain to the Ross of Mull and Fionnphort, the crossing point to one of the Christian world's most famous small islands – Iona.

At this point I am tempted to include a photo with the caption 'Somewhere on Mull', taken many years ago. Actually I am fairly sure it is a mountain view in Glen More, about 9 miles from Craignure, and I love it for showing the island in a very different mood from that of the typical sun-drenched tourist brochure – soothing, unhurried, unpeopled, and bathed in a misty morning light. Mull has a very small population for an island of its size, and offers plenty to walkers and seekers of solitude.

Yet there is a sad side to such beauty, because the island's inhabitants, like those of Skye, suffered greatly from 19th-century clearances. Writing about Mull in 1818, a resident of nearby Coll reported

a slow ruthless policy of evictions. All the lairds were in it with few exceptions. Poor people were compelled to work on roads, drains, walls, fences and piers for a scanty allowance of oatmeal and other victuals, much of it sent gratis from America to be distributed among the poor, just to keep them alive. This was exploited by the lairds to their advantage.

Left: Main Street, Tobermory, overlooking Telford's harbour (Geograph/Jo Turner).

Right: Somewhere on Mull (Paul A. Lynn).

He was referring to the first wave of evictions between about 1780 and 1820, when lairds moved tenants to coastal settlements, replaced them with sheep, and 'encouraged' the people to take up fishing, kelping, and other activities alien to their traditional way of life. The second wave, which began in the 1840s and lasted for nearly 50 years, was far more extensive and caused mass emigration, especially to North America. In many cases the lairds sold out to new landowners who had even less interest in their tenants' welfare and established 'sporting estates' to entertain the wealthy. Perhaps we should set Mull's present gift of solitude against the pathetic little rectangles of ruined cottage walls so often visible among its straths and glens.

The first time I reached Fionnphort ferry terminal, I recall having to buy tickets in a depressingly tatty building and thinking that little Iona, just a mile across the water and internationally famous, deserved something far better. The situation has now been remedied (although the crowds have increased greatly in the past 20 years). Anyway it is all worthwhile, especially if you can avoid the main tourist season.

Iona, just 3 miles long and 1½ wide, is steeped in history and tradition. St Columba, a prince of Ireland and follower of St Patrick, landed here in AD563, not as a missionary as is commonly supposed, but to do penance for a violent act of bloodshed. He found an island steeped in pagan worship, so there was plenty to stir his missionary instincts. Its soil would eventually become the resting place of 48 Scottish kings. Columba and his disciples went on to spread Christianity further afield, converting the king of the Picts at Inverness, and crossing over to mainland Europe, even as far as the Rhine valley. Today Iona Abbey attracts thousands of pilgrims and visitors, and can claim to be one of the finest ecclesiastical buildings in the Highlands and Islands of Scotland.

You may be thinking that Iona is simply another diversion from our main Telford itinerary – and you would be right, except for one significant detail. For although his roadbuilding programme could hardly be expected to encompass such a miniature island, Iona was granted one of Telford's churches, largely financed by the British government and built in the

Left: On a visit to Iona Abbey (Paul A. Lynn).

Right: The Telford church on Iona (Geograph/Ian Capper).

1820s. Standing back from the main road to the abbey, this one is still in fine condition and acts as Iona's parish church. Nearby is MacLean's Cross, a tall free-standing cross probably erected in the 1400s as a prayer station for visiting pilgrims.

It is time to return to Mull and plan our onward journey to the Isle of Islay, about 40 miles to the south, where Thomas Telford had considerable influence. The most attractive option is surely to do some island-hopping: from Craignure on Mull to Oban on the mainland; and from Oban to Port Askaig on Islay, via the far smaller island of Colonsay. Ferry timetables at the ready!

Islay and Jura

Islay, the third-largest Inner Hebridean island, has a very different character from Skye and Mull. For a start there are no Munros or Corbetts (although its highest mountain apparently ranks as a Marilyn: under 2,000 feet but with at least 492 feet clearance on all sides). It is also more agricultural, a natural place for a good crop of barley, the essential ingredient of Scotch whisky. The harbour at Port Askaig is little more than a dent in the shoreline along the narrow Sound of Islay which divides Islay from Jura and twice daily exhibits one of Europe's fastest tidal streams.

Leaving the harbour at Port Askaig we climb steeply uphill on the A846. Within a couple of miles there are two right turns onto single-track roads leading to whisky distilleries on Islay's shoreline: Caol Ila and Bunnahabhain, two of the brands for which Islay is world-famous among lovers of a wee dram. About a mile beyond the village of Keills there is yet another right turn, again onto a single-track road, but this time to Finlaggan, the administrative centre of the Lordship of the Isles during the 13th to 15th centuries – one of Scotland's most historic sites. And finally, just beyond the village of Ballygrant, on the left and more or less hidden in trees, is our first Islay item of Telford interest: Kilmeny Parish Church. The four locations make an interesting day's excursion.

Some years ago, after visiting the whisky distilleries and Finlaggan by design, we came across Kilmeny Church by accident. A sign at the roadside advertised a church, so we wandered up the driveway. A gardener with a key in his pocket kindly opened the building up for us. He had heard of Telford, but was vague about his churches. Anyway we had seen enough to convince us that this was one of them – and a well-kept one at that. Most visitors to the island presumably pass it by unnoticed because of the enveloping trees.

About 5 miles beyond Kilmeny we come to the village of Bridgend, where the A846 turns south towards Bowmore (another distillery), Port Ellen (another ferry terminal), and the south coast of Islay (including the Laphroaig, Lagavulin, and Ardbeg distilleries). Bowmore is an especially attractive small town, with a broad main street leading up to its famous Round Church, built when Thomas Telford was just ten years old. It used to be claimed that the circular design avoided sharp corners in which the devil could hide.

Above: The harbour at Port Askaig (Paul A. Lynn).

Above, right: Caol Ila distillery overlooks the Sound of Islay (Geograph/John Allan).

Right: Kilmeny Parish Church (Geograph/Andrew Abbott).

Below: Main Street and the Round Church, Bowmore (Geograph/M.J. Richardson).

The Still Room at Bowmore Distillery (Paul A. Lynn).

So far we have been travelling on roads untouched by Telford, but the situation is easily rescued by going back to Bridgend and taking a left turn onto the A847. The 15-mile coastal stretch to Portnahaven was part of his infrastructure programme; and as luck would have it we pass fairly close to a modern distillery (Kilchoman), even closer to a historic one (Bruichladdich), and through the delightful village of Port Charlotte. It is an excursion not to be missed (even though we must return to Bridgend the same way).

Telford's road opened up the Rhinns of Islay peninsula and the far southwest of Islay at Portnahaven. Attractive Shore Street is lined with what look like Telford-inspired cottages, and there is a fine Telford church dated 1828. The neighbouring village of Port Wemyss, built a few years later, was founded by the local laird, Walter Campbell of Islay, to rehouse some of the people he was busy clearing from inland areas of the island. Port Wemyss never had a church of its own, so residents walked the short distance to Portnahaven and entered its church by one of the doorways, leaving the other for 'true locals'.

The small island of Orsay, separated from Portnahaven and Port Wemyss by 200 yards of swirling tidal water, is host to one of the most important buildings on the west coast of Scotland – the Rhinns of Islay lighthouse, completed by Robert Stevenson in 1825. The two villages must have been hives of activity in the early 1800s, and I imagine the plans for Telford's road, Stevenson's lighthouse, and Campbell's village had quite a lot to do with one another.

We now retrace our route 37 miles through Bridgend to Port Askaig, and await the doughty little ferry that crosses the Sound of Islay to Feolin on the Isle of Jura. This will be the last port of call on our Telford tour of the Inner Hebrides.

Left: Portnahaven (Geograph/Paul Birrell).

Right: Telford's church at Portnahaven (Geograph/Andrew Wood).

Jura, about 30 miles long by 7 miles wide, is the wildest and emptiest of Scotland's large inhabited islands. The present human population of about 200 is complemented by 30 times as many deer, and overlooked by three famous mountains, the Paps of Jura (the highest a Corbett, the others nearly so). Their steep quartzite slopes and appropriately shaped summits are often visible from Skye and Mull, and they dominate the view from Port Askaig.

The Feolin Ferry must battle twice a day against powerful tidal races in the Sound of Islay, and its course can seem strangely curvaceous to the uninitiated. But it reliably reaches the landing ramp at Feolin and the start of Jura's only road. The A846 runs 8 miles to Craighouse, Jura's only village, and a further 16 miles up the east coast to Lagg and the tiny settlement of Ardlussa. In former times drovers took thousands of cattle reared on Islay via Feolin to the harbour at Lagg, from where they were ferried across the Sound of Jura for sale on the mainland.

As far as I know, Jura's A846 is the only lengthy, single-track, Class A cul-de-sac in Britain, with grass growing intermittently in the middle. I have sometimes driven the 8 miles from Feolin to Craighouse without meeting a single vehicle, and felt wonderfully free of life's

Left: Rhinns of Islay Lighthouse (Geograph/G. Laird).

Right: The Paps of Jura seen across the Sound of Islay (Geograph/Ian Andrews).

Left: The Feolin Ferry approaches Jura from Port Askaig. In the background is the Caol Ila distillery on Islay (Geograph/M.J. Richardson).

Right: Tarmac with grass: the A846 between Feolin and Craighouse (Geograph/Richard Webb).

concerns. Initially the road hugs the shoreline of the Sound of Islay, then curves gently inland towards Craighouse and its attractions – Jura's whisky distillery (founded in 1810, restored and expanded since the 1960s); its close neighbour the Jura Hotel, with extensive views over Small Isles Bay; the Jura Stores, third-largest building in Craighouse; the village hall, cream-painted with a bright red roof; and (at the time of writing) a welcoming bistro restaurant.

Jura's old pier, still in good order, was built to serve the distillery. I remember strolling along its impressive stonework many years ago, although at that stage I had no idea it had anything to do with Thomas Telford. The much longer new pier offers a panoramic view of the village and surrounding countryside.

Left: The old pier at Craighouse (Geograph/Chris Downer).

Right: Craighouse seen from the new pier: at centre left, the distillery and hotel; at far right, the village stores, village hall, and old pier (Geograph/M.J. Richardson).

*The A846 approaches
the harbour at Lagg Bay
(Geograph/M.J. Richardson).*

Telford was also involved with Lagg Harbour, 9 miles up the coast from Craighouse. In the old days the ferry crossing from Lagg to Keills in Knapdale was the normal route for exporting cattle to the mainland. An application for a parliamentary road to link the ferries at Feolin and Lagg was made in 1804. Two dry stone rubble piers were completed at Lagg by 1810, and certified by Telford's inspector two years later. The ferry then became the official mail carrier to and from Jura and Islay. Little Lagg, which seems so tranquil today was, like Portnahaven on Islay, a hive of 19th-century activity.

Having come this far up the A846, you may wish to complete the job by continuing 7 miles to the small coastal settlement of Ardlussa. Beyond is a private track leading towards Barnhill, the farmhouse which sheltered George Orwell while he wrote his novel *1984* in 1948. And if you are game for a long walk there and back, you will come to the northern tip of Jura and overlook the narrow Gulf of Corryvreckan. The roaring of its notorious whirlpool can often be heard miles away. Not, you might think, ideal waters for a small dinghy; but George Orwell risked it, almost drowning himself and several others in the process.

Telford's highland churches

In the early 19th century, the British government decided that the provision of churches no longer matched the spiritual and religious needs of the population. The Highlands and Islands of Scotland presented a special problem because many of their parishes were large areas of rough mountain country. Parishioners, including those cleared from their crofts and rehoused in new coastal villages, often lived so far from a parish church that they could not attend worship regularly (even if they wished to). In addition, competing religious sects and denominations, including the Roman Catholic, were seen as undesirable by the London government.

Following the Napoleonic wars, Parliament made available £1,000,000 in 1819, and a further £500,000 in 1824, for the building of churches and chapels for the Church of England, as an expression of gratitude to God for victory. The resultant 214 Commissioners' Churches were built or refurbished, one of which cost £77,000.

A similar proposal to provide £200,000 for the Church of Scotland was delayed for years by political difficulties and obstructions, and when an amended Bill was eventually passed in 1824, it provided just £50,000 for the whole of the Highlands. About 30 churches, often referred to in Scotland as Parliamentary Kirks, were to be built, some with manses for their ministers – but no more than £1,500 was to be spent on any one site. A similar Bill for the Lowlands failed altogether in 1825. The majority of Scottish parishes, and parishioners, received nothing.

The task of selecting the sites and overseeing the work was entrusted to the Commissioners for Building Highland Roads and Bridges, and in particular to Thomas Telford. It would be his last major project in the Highlands and Islands. The Bill stipulated that heritors (the principal local landowners in a parish) could apply for a new church to be built on their land, and in 1825/6 the commissioners considered 96 applications. Eventually, and not without difficulty, 32 sites were chosen, and the building programme was completed by about 1830.

Inevitably some of these Parliamentary Kirks have not survived the intervening years; others are derelict; and of the remainder, some have been converted to secular use. The

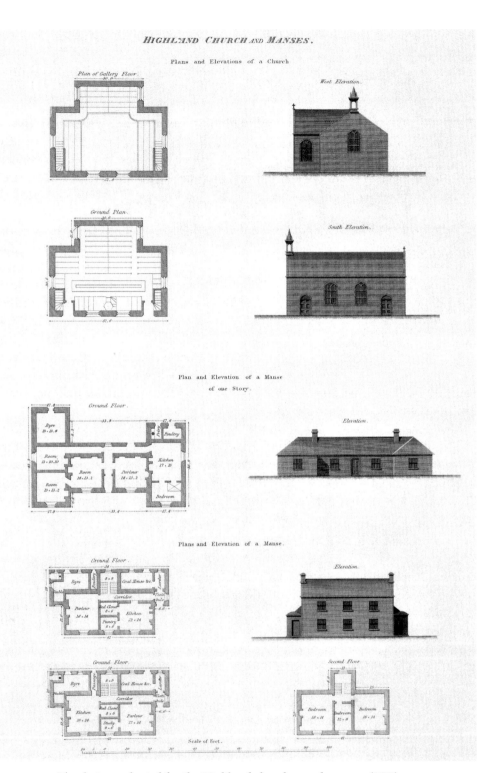

The designs adopted for the Highland churches and manses (ICE).

Locations of 20 surviving Parliamentary Kirk buildings. The five shown dark green (Berriedale in Caithness; Hallin on Skye; Iona; and Kilmeny and Portnahaven on Islay) have been mentioned previously).

accompanying map shows the locations of 20 surviving kirk buildings in the Highlands and Islands, in various states of preservation. Some are still used as parish churches.

Telford asked three of his surveyors, William Thomson, James Smith, and Joseph Mitchell, to prepare detailed designs for a church and two sizes of manse. All had to be within budget and 'particularly calculated to resist a stormy climate'. After various amendments, he approved Thomson's plan for the church: a simple basic rectangle, with various options to suit local circumstances including an extension at the back. The shape and positioning of doors, windows, and a small belfry were specified, and landowners could add internal lofts or galleries at their own expense. Standardised windows were supplied, ready to fit, by James Abernethy of Aberdeen. Smith's design for a single-storey manse, and Mitchell's for a two-storey version, were also adopted.

Building commenced in 1826, and all but four of the sites were completed within the budget of £1,500. In most cases *quoad sacra* parishes (that is, ecclesiastical but not civil), were established around the new churches, giving their ministers clear boundaries to work with, but even so disagreements with the ministers of the pre-existing parishes sometimes occurred. The new ministers received an annual stipend of £120, from which they were expected to maintain the manse, leaving the heritor to maintain the church.

The Telford church at Kilmeny, Isle of Islay (Geograph/M.J. Richardson).

Scotland Revived

We have already met Scotland in Need in the years of Thomas Telford's childhood: its poverty compared with England; the almost complete lack of proper roads and bridges, especially in the West Highlands and the Islands; clans downtrodden after Bonny Prince Charlie's rebellion of 1745; the gathering storm clouds of heartless and hasty evictions by certain landowners and clan chiefs who preferred sheep to Highlanders.

So how about Scotland Revived? What did Telford himself feel about the situation he inherited, the responsibilities he shouldered, the prodigious network of roads and bridges he designed and supervised? I will start with some of his own words:

> In these works, and in the Caledonian Canal, about three thousand two hundred men have been annually employed. At first, they could scarcely work at all: they were totally unacquainted with labour; they could not use the tools. They have since become excellent labourers, and of the above number we consider about one-fourth left us annually, taught to work. These undertakings may, indeed, be regarded in the light of a working academy, from which eight hundred men have annually gone forth improved workmen …
>
> Since these roads were made accessible, wheelwrights and cartwrights have been established, the plough has been introduced, and improved tools and utensils are generally used. The plough was not previously employed; in the interior and mountainous parts they used crooked sticks, with iron on them, drawn or pushed along. The moral habits of the great masses of the working classes are changed; they see that they may depend on their own exertions for support: this goes on silently, and is scarcely perceived until apparent by the results. I consider these improvements among the greatest blessings ever conferred on any country. About two hundred thousand pounds has been granted in fifteen years. It has been the means of advancing the country at least a century.

Telford can hardly be accused of mincing his words. Although he professed a lifelong love of his native land, he was quite prepared to criticise what he perceived as wilful backwardness. A particular scandal in his eyes was the use of 'crooked sticks, with iron on them' for tillage, rather than horse-drawn ploughs. As Samuel Smiles confirmed:

> The plough had not yet penetrated into the Highlands; an instrument called the Cas-Chrom – literally the "crooked foot" – the use of which had been forgotten for hundreds of years in any other country in Europe – was almost the only tool employed in tillage in those parts of the Highlands which were separated by almost impassable mountains from the rest of the United Kingdom.

In fact the cas-chrom remained far from banished in Telford's, or even Smiles's, lifetime – for example on the Isle of Skye, where it survived well into the 20th century.

Smiles summarised Telford's Highland legacy some 50 years after the events, with all the benefits of hindsight and none of the risks of instant praise:

> Thus, in the course of eighteen years, 920 miles of capital roads, connected together by no fewer than 1200 bridges, were added to the road communications of the Highlands, at an expense defrayed partly by the localities immediately benefited, and partly by the nation. The effects of these twenty years' operations were such as follow the making of roads everywhere – development of industry and increase of civilisation. In no districts were the benefits derived from them more marked than in the remote northern counties of Sutherland and Caithness. The first stage-coaches that ran northward from Perth to Inverness were tried in 1806, and became regularly established in 1811; and by the year 1820 no fewer than forty arrived at the latter town in the course of every week, and the same number departed from it. Others were established in various directions through the Highlands, which were rendered as accessible as an English county.
>
> Agriculture made rapid progress. The use of carts became practicable, and manure was no longer carried to the field on women's backs. Sloth and idleness gradually disappeared before the energy, activity, and industry which were called into life by the improved communications. Better built cottages took the place of the old mud biggins with holes in their roofs to let out the smoke. The pigs and cattle were treated to a separate table. The dunghill was turned to the outside of the house. Tartan tatters gave place to the produce of Manchester and Glasgow looms; and very soon few young persons were to be found who could not both read and write English.

From a personal point of view one of the most remarkable aspects of Telford's Highland programme was its coincidence with his other huge projects. Completion of the Pontcysyllte Aqueduct in North Wales in 1805 brought him national and international recognition. In 1808, while inspecting works on the Caledonian Canal, he received a letter from the King of Sweden asking for help with another major ship canal (the Gotha Canal, mentioned earlier, between the North Sea and the Baltic), a project that would run for 24 years. In 1815 the British government asked him to recommend an improved mail route between London and Ireland; the work on his road to Holyhead (now the A5) continued for 15 years and included the world-famous Menai Suspension Bridge linking North Wales to Anglesey. His advice as a consulting engineer was constantly sought on other projects, including many canals and roads. For example, floundering plans for a Gloucester & Sharpness ship canal and for a Thames & Medway canal landed on his plate. He was also asked to advise on improvements to the Great North Road from London to Edinburgh, and on an alternative route to Ireland via the new steam packet service from Milford Haven to Waterford. And so it continued: he took on many projects at home and abroad even while heavily engaged in the Highlands and Islands.

How could one man find the time and energy to take all this responsibility; surely he was grossly overworked, and sometimes in a frenzy of worry? The answer seems to lie in his fundamental nature, which went right back to the genes he had inherited and the Laughing Tam personality of his youth in Eskdale. As author Tom Rolt explained in his biography of Telford, published in 1958:

> Work for Telford was no obsession and consequently it never robbed him of his humanity; never consumed him. Few men of his generation could have boasted a wider or more varied circle of friends from Members of Parliament, Government officials, great landowners, men of science and fellow engineers to labourers, working craftsmen and country innkeepers. On his Highland journey with Telford, Southey repeatedly remarked on the warmth of the welcome they received at every stopping place. Everywhere Telford appeared to be well known and so well loved that his coming was an important event. With most of the innkeepers and "locals" whom they encountered in the Highlands he appeared to be on Christian or nickname terms. He would chaff them, remember the appropriate stock jokes and afterwards regale Southey with local information, history and gossip.

Such was the character of the outer man. The inner one, appreciated by Robert Southey as a fellow poet, was nourished by Telford's enduring love of home, and especially the hills of Eskdale, which eclipsed the attractions of ambition and wealth. The final stanza of his reworked poem 'Eskdale' speaks volumes:

Yet still, one Voice, while fond Remembrance stays,
One feeble Voice, shall celebrate thy praise,
Shall tell thy Sons that, wheresoe'er they roam,
The Hermit Peace hath built her Cell at home;
Tell them, Ambition's wreath, and Fortune's gain,
But ill supply the Pleasures of the Plain;
Teach their young Hearts thy simple Charms to Prize,
To love their native Hills, and bless their native Skies.

From 1801, when Telford had received a letter from the government requesting a survey and report on Highland communications, he inevitably found himself immersed in politics, so needed a London base. He reserved several rooms at the Salopian Coffee House near Charing Cross, a venue that attracted him because Salop was the old name for Shropshire, where he had spent so many formative years. The rooms became his professional HQ, a regular meeting place for fellow engineers, assistants, and visitors from abroad.

The arrangement lasted for 21 years, by which time he was 64 years old – famous and well able to afford a house of his own. Then, after all the years of frenetic travel – not least in the Highlands and Islands of Scotland – Telford found a permanent home, 24 Abingdon Street, opposite the Houses of Parliament. Not that he kept it entirely for himself: rooms were set aside for his younger assistants and pupils, including a nephew of his old blind friend, Andrew Little, and Joseph Mitchell, son of his 'Tartar', John Mitchell. Although Telford had no desire to enter London's smart society, he gave dinner parties for friends and colleagues, including Robert Southey, and kept guests amused with steady streams of reminiscences, jokes, and personal anecdotes.

Another reason why Telford settled down at a secure London base was his close involvement with the Institution of Civil Engineers (ICE), the world's first professional engineering body. In 1818 his assistant, Henry Palmer, and other young engineers started pressing for something more practical than the Society of Civil Engineers, which had been founded in 1771 by John Smeaton and which they felt had become just a private dining club. An inaugural meeting of the ICE was held at Kendal's Coffee House in Fleet Street, but progress was slow until 1820, when Telford agreed to become the institution's first president. He was determined to make it a fount of engineering knowledge and achievement that would promote civil engineering as a profession. His determination paid off: it was granted a Royal Charter in 1828, and Telford remained president until the end of his life.

By the early 1830s, however, Telford's professional life was drawing to a close. Increasing deafness made him feel uncomfortable in society, even among friends. The last major project on which he was consulted – by the Duke of Wellington, at that time Lord Warden of the Cinque Ports – was the 1834 improvement of Dover Harbour to guard against a cross-Channel invasion. A few months later Telford was laid up by 'bilious derangement of a serious character' and on 2 September 1834, at the advanced age of 77, he passed away.

Typically Telford had directed that his remains should be laid without ceremony in the graveyard of his parish church, St. Margaret's Westminster. But Westminster Abbey was to offer him a final resting place, and space for a larger-than-life marble statue showing him with a pair of dividers in one hand and two engineering books in the other. The inscription reads:

THOMAS TELFORD.
PRESIDENT OF THE INSTITUTION
OF CIVIL ENGINEERS:
BORN AT GLENDINNING IN ESKDALE
DUMFRIES-SHIRE IN MDCCLVII.
DIED IN LONDON MDCCCXXXIV.

THIS MARBLE HAS BEEN ERECTED NEAR THE SPOT
WHERE HIS REMAINS ARE DEPOSITED
BY THE FRIENDS WHO REVERED HIS VIRTUES:
BUT HIS NOBLEST MONUMENTS ARE TO BE FOUND
AMONGST THE GREAT PUBLIC WORKS OF THIS COUNTRY.
THE ORPHAN SON OF A SHEPHERD, SELF EDUCATED,
HE RAISED HIMSELF BY HIS EXTRAORDINARY
TALENTS AND INTEGRITY FROM THE HUMBLE CONDITION
OF AN OPERATIVE MASON, AND BECAME ONE OF
THE MOST EMINENT CIVIL ENGINEERS OF THE AGE.

So this man of the Scottish Enlightenment, a world-famous engineer whose formal education had started and ended in a Lowland parish school, was finally honoured by the nation.

And how might those of us who live far from the metropolis remember him? Perhaps by coming full circle on this story with a journey to Bentpath and its remarkable library, there recalling the love of Eskdale that inspired him throughout a long and fruitful life, and his extraordinary mission in the Highlands and Islands –- a heartfelt contribution to 'Scotland Revived'.

The library and Telford memorial at Bentpath (Geograph/ Walter Baxter).

Index

1. People

2. Scottish placenames (first mention)